John Norton Loughborough

**Hand Book of Health**

John Norton Loughborough

**Hand Book of Health**

ISBN/EAN: 9783337371029

Printed in Europe, USA, Canada, Australia, Japan

Cover: Foto ©berggeist007 / pixelio.de

More available books at **www.hansebooks.com**

# HAND BOOK OF HEALTH;

OR, A BRIEF TREATISE ON

# Physiology and Hygiene,

COMPRISING

PRACTICAL INSTRUCTION ON THE STRUCTURE AND
FUNCTIONS OF THE HUMAN SYSTEM, AND
RULES FOR THE PRESERVATION
OF THE HEALTH.

## BY J. N. LOUGHBOROUGH.

"I will praise Thee; for I am fearfully and wonderfully made."—DAVID.

STEAM PRESS
OF THE SEVENTH-DAY ADVENTIST PUBLISHING ASSOCIATION,
BATTLE CREEK, MICH.

1868.

# PREFACE.

In presenting the following pages to the public, the compiler does not do it because no works have been written upon this subject, nor because his researches in the study of the human system have been greater than that of others. Many, and able works have been written, but the best of them are quite large, so that many will not take time to read them, even if their means would permit their owning them. In many treatises on physiology there are thoughts presented, important indeed to the physician, but which fail to interest the common reader.

Seeing a necessity for a work which should, in as brief a manner as possible, give some general ideas of the structure, functions, and care of the human system, I have been induced to prepare the following pages; not so much as an original production as a compilation of general ideas from different authors, simplifying, in some instances, their language to the capacity of the common reader.

The author would here acknowledge the use of the experience, researches, and writings of such men as GRAHAM, TRALL, LAMBERT, HITCHCOCK, MENDENHALL, TAYLOR, and many others, and commend their writings, especially *Graham's Lectures on the Science of Human Life*, and *Dr. R. T. Trall's Hydropathic Encyclopedia*, to the perusal of those who are able, and who wish to go into a deeper investigation of these matters.

## PREFACE.

In arranging this book, the form of questions and answers has been chosen, not because it is especially designed for a school book,—although it might be thus used, by still other questions being propounded by the teacher,—but to impress more forcibly the mind of the reader.

In the explanations of the few cuts that are introduced, I have tried to give the use of the different parts illustrated, rather than perplex the mind with technical terms. Now, that the task is completed, these pages are commended to your candid perusal with the hope that they may prove a source of good in correcting wrong habits, inculcating temperance, and a regard to the sacred laws implanted in our own natures, and thus bless the minds and bodies of thousands.   J. N. LOUGHBOROUGH.

*Battle Creek, Mich., Jan.* 1, 1868.

---

## ENDORSEMENT.

HAVING carefully examined the manuscript of this work, I can cheerfully recommend it to the confidence of the public as being well adapted to the wants of the common people, and in accordance with the recognized principles of physiology, and of hygienic medication. It should be in every family, and read and studied by old and young. The chapter on the "Nervous System" in my opinion is worth more than the price of the book.

H. S. LAY. M. D.
*Health Reform Institute, Battle Creek, Mich., Jan.* 1, 1868.

# TABLE OF CONTENTS.

### CHAPTER ONE.
#### PHYSIOLOGY AND HYGIENE.
*Pages 1—18.*

Explanation of the subject.—General remarks on the Human System, and the importance of understanding it, and knowing how to properly care for it.

### CHAPTER TWO.
#### THE HUMAN FRAME.
*Pages 19—42.*

The bones of the body: their construction, number, nature, power, position, and use.—The joints, ligaments, synovia, &c., &c.

### CHAPTER THREE.
#### THE MUSCULAR SYSTEM.
*Pages 43—61.*

The muscles: their construction, number, power and manner of action.—Muscles of the head and face, neck, back, chest, limbs, &c.—Pairs of muscles.—Muscles of the alimentary canal.—Disadvantageous action of muscles.—Rapidity of muscular motion.—Care of the muscles.—Muscular exercises.—Fascie.

### CHAPTER FOUR.
#### THE CIRCULATION OF THE BLOOD.
*Pages 61—81.*

The heart described.—Its auricles and ventricles, and their action.—Valves of the heart.—Course of the blood through the heart.—The time the heart rests.—Times of its beating.—Its power and capacity.—Heart illustrated.—The arteries: their number and capacity.—Their position.—Their wounds; how known, how treated.—Their origin, and number.—Connection between arteries, and why.—Pulmonary arteries.—Capillaries: their rise, number, and use.—Their position.—Venous system.—Structure of the veins.—Their origin, and course.—Vein-valves, and their use.—Three classes of veins.—Forces of circulation in veins.—Pulmonary veins.—Portal system.—Blood: its color, and quantity in the body.—Time of its circuit through the body, and amount passing through the heart per hour.—What will promote a good circulation.—Clothing, proper food, &c.—Disturbing causes in the circulation.

## CHAPTER FIVE.

### The Lymphatics.

*Pages 82—87.*

The office of the lymphatics.—Their origin, and construction.—Lymphatic glands.—Lacteals.—Difference between lymphatics and lacteals.—Different kinds of lymphatics.—Radicles.—Chyle.—Lymph.—General views of absorption in the system.

## CHAPTER SIX.

### The Nervous System.

*Pages 88—124.*

Nerves.—Two nervous systems in living animal bodies: organic, and nerves of animal life.—Structure of nerves, and the size of component parts of nerves.—Center of organic nervous system.—The solar plexus.—Ganglia.—Suspension of the action of organic and animal nerves: the effect.—Distribution of animal nerves.—Nourishment of nerves.—Illustration of sympathetic, or organic nerves.—Important relation of the stomach to the nervous system.—The cerebro-spinal, or animal nervous system.—Its center, the brain.

—Structure of the brain: its size, divisions, and coverings.—Cerebrum.—Cerebellum.—Cerebro-spinal nerves: their origin.—Spinal marrow.—Medulla oblongata.—Cranial nerves.—Arbor vitæ.—Ganglia of the brain.—Nine pairs of nerves.—Thirty-one pairs of spinal nerves.—Action of the brain.—Injuries of the brain: the effect.—Inactive in profound sleep.—Sleep: proper amount; how taken; in what rooms; best time for sleep, &c.—Exercise of the brain: result.—Phrenological arrangement of the brain.—The mind of man.—Faculties, and propensities.—True happiness.—Effect of mind on body, and of the body on the mind.—Pneumo-gastric, or lung-and-stomach nerve.—Its importance.—Effects of the mind on the stomach, and of the stomach on the mind.—Disease of the nervous system.

## CHAPTER SEVEN.
### ORGANS OF THE EXTERNAL SENSES.
*Pages* 125—142.

Organ of smell: its structure, action, and care.—The eye: its structure, and action.—The medium of sight.—Seeing illustrated.—Care of the eye.—The ear: its structure, action, benefit.—Diseases of the ear, and proper care.—The organ of taste, the tongue: its structure, and action.—What a healthy taste is.—Sense of touch.—The skin described.—Its derma, epiderma, sweat glands, hair follicles, nails, &c.—Action of the skin as a depurating and breathing organ.—Care of the skin.—Bathing.—Instructions about bathing.—Colds: how produced, how cured.—The clothing as related to the care of the skin.

## CHAPTER EIGHT.
### THE VISCERA.
*Pages* 143—178.

Three cavities of the body: crano-spinal, thorax, and abdomen.—Organ of the voice, the larynx described.—How the voice is produced.—Training the voice.—The wind-pipe:

its description, and action.—The lungs: their action.—The pleura.—Breathing: how accomplished.—Capillaries of the lungs.—Digestion of air in the lungs.—Pure air essential to health of lungs.—Purification of the blood in the lungs.—Freedom for the lungs.—Effects of impure air on the lungs.— Ventilation of rooms. — Inflammation of lungs: how treated.—The peritoneum.—Great omentum and mesentery.—Alimentary canal.—The mouth, and salivary glands.—Effects of improper mastication.—Drinking with food: its effect.—The œsophagus.—The stomach: its action.—Duodenum, or second stomach. — Pancreas. — Liver.—Gall bladder.—Jejunum.—Ileum.—Large intestine.—Rectum.—Mesocolon. — Caul. — Spleen. — Kidneys. — Supra-renal capsules.

## CHAPTER NINE.
### Diet, or Proper Food and Drink.
*Pages* 179—198.

Elements of the body.—Elements of food.—Law of adaptation.—Meat.—Alcohol, and stimulants.—Comparative nutriment of meat and vegetable food.—Disease of flesh meats.—Butter, milk, cream, cheese, flesh soups, fish, eggs, acids, vinegar.—Proper food.—General instruction relative to eating.—Food of children.—Pure soft water the most wholesome drink.

## CHAPTER TEN.
### Miscellaneous Items.
*Pages* 198—205.

Animal heat: how produced.—Exercise.—General instructions relative to exercise.—Disease.—Cause of disease.—Cure of disease.—Medicines: their effect.—Passions.—Secret youthful vices.

Index, . . . . . . 206–213.
Glossary, . . . . . 214–227.

# PHYSIOLOGY AND HYGIENE.

## Chapter One.

### EXPLANATION OF THE SUBJECT.

1. What is the meaning of the word *Physiology?*

This word is formed from two Greek words, *phusis*, nature, and *lego*, to discourse. It means a description of Nature.

2. Of what does Physiology treat?

It treats upon the purposes, uses, functions, actions, properties, results, and relations, of the various parts of the living body, in its healthy or *normal* condition. In other words, it is a description of organized or animate Nature, such as trees, plants, brute animals, and man.

3. Into what branches is Physiology divided?

Into Vegetable, Animal, and Human. When treating of plants and trees, it is *Vegetable Physiology.* When treating of brute animals, it is *Animal Physiology.* *Human Physiology*, or that relating particularly to man, is that which we shall treat upon in the following pages.

4. What is *Hygiene?*

The meaning of the word Hygiene, is health. As a study, Hygiene treats upon what will improve and preserve health, what will impair and destroy health, and what is best calculated to produce a healthy condition of the body.

5. Does not a work of this kind embrace the subject of *Anatomy?*

Anatomy treats upon the structure of living things, such as their color, size, form, surface, &c.

6. Will this work, then, treat upon Anatomy?

It will treat upon it only so far as it is connected with Physiology and Hygiene. We design to introduce, in each of these branches, only that which is of use to all classes of persons. The work then must contain something of Anatomy, more of Physiology, and much of Hygiene. These branches are so intimately related to each other, that we do not propose to separate them in this work, but to treat upon the structure, functions, and care of the different parts of the body as we pass along.

7. What does man discover concerning his relation to surrounding objects, as he enters upon the study of himself?

He discovers that his senses, feelings, and faculties, relate him to the whole universe. His well-being certainly demands that such relation should be the most harmonious. The world appears to him full of beauty, and he has eyes adapted to see it, and faculties just fitted to enjoy it. His ears are wonderfully adapted to all sounds, and their harmonious combination in music affords the most pleasing sensation. His sense of smell is related to a thousand delightful odors. His taste finds exquisite gratification from the aliments best adapted to supply the waste of his system. His sense of touch, variously modified in many organs of the body, gives him a world of delight.

8. What benefit can we derive from the study of ourselves?

It will discover to us, that man, in his nature and faculties, capabilities and conditions, and in his relations to the world in which he exists, is one of the most interesting and important subjects which the human mind has power and compass to investigate. This study also displays to us the wonderful wisdom of God in our formation, and teaches us how we can secure the greatest amount of happiness here, by showing us that true enjoyment in this world can only be secured by the greatest possible freedom from sickness and pain.

9. How can such a study teach us the art of happiness?

By teaching us what kind of food, air, and habits of life, will tend to make us sick; what kinds will preserve health; and how we can obtain cheerfulness, and freedom from that health-destroying disease, despondency of mind.

10. What does Physiology discover to us concerning the mind?

It discovers to us, that to properly understand and care for the mind, it is necessary to ascertain how far the mind is connected with the body; to what extent it is affected by the conditions of the body; and then, again, on what depend those conditions of the body which affect the mind. In order to this, the body itself must be understood in its animal and organic nature, its physical and vital properties and laws, and its physiological actions and affections. This will show us that mind and body are so closely connected that one is affected by the other. So that we cannot habitually possess lively and correct moral feelings, or a sound mind, unless we so live as to preserve a sound body; for true happiness may be properly defined as health of body and health of mind.

11. How, then, does Physiology treat of the human body?

As a system, composed of sub-systems, all being called the *human system*. The whole composed of dependent parts, acting upon each other, but all working together harmoniously.

12. What, then, does man's well-being here require?

It requires a knowledge of himself, both mentally and physically, and also such a relation of himself to the elements of the external world, and such adaptation of these elements to his own condition, that they may, through the body, act favorably upon the mind.

13. What is required in properly treating upon the care of the human body?

To properly treat upon the care of the human body, it is necessary to take into view, and thoroughly investigate, the nature, conditions, and relations of man; to understand the modifying influences of the mind and morals upon the health and morbid sensibilities and sympathies of the system. Man finds himself upon the stage of life, surrounded by innumerable influences, acted upon at every point, and he is continually conscious, not only of his own existence and the action of surrounding influences, but of an unceasing desire for happiness. This desire itself is a living proof that our benevolent Creator has fitted us for happiness, not only in a future state, but here; and he has adapted everything within us and around us to answer this desire, in the fulfillment of those laws of life, health, and happiness, which He, in wisdom and in goodness, has established in the constitutional nature of things.

14. What are the faculties of all living bodies?

All living bodies possess those faculties by which their nourishment and growth are effected, and their temperature regulated. The little acorn placed in a genial soil, other circumstances being favorable, is excited to action by virtue of its own vitality. It puts forth its roots, twigs, branches and leaves, till it becomes a giant oak. All the vital operations of the tree are maintained till the vital property is worn out or destroyed, when its death ensues. The tree by nature is fixed to the spot from whence it sprang,—unconscious of its being, without any organs of external perception, or voluntary motion. So far as the vital operations are considered by which chyme, chyle and blood are produced, the blood circulated, the body in all its parts nourished, and its growth effected, its temperature regulated, and all the other functions of organic life sustained, man is as destitute of animal consciousness as the oak. But in man there are two classes of functions. Besides the class already mentioned, concerned in the growth and sustenance of the body, there is a secondary class, which consists of those functions which minister to the wants of the primary class. This class is established with special reference to the relation existing between those internal wants and the external supplies, and general external relations of the body. This second class of functions is peculiar to animal bodies.

15. What are the powers of the vital economy?

The vital economy seems to possess the power of supplying from the common and ordinary current of blood, without any known variation in the food from which it is formed, a large increase of appropriate nourishment for particular structures, and at the same time regularly sustaining the general function of nutrition in every part and substance of the system. From the same chyle various substances are produced, opposite in their qualities, and composed of essentially different elements. The flesh of the rattlesnake is eaten by many as a great luxury, and its blood may be put upon a fresh wound with perfect safety; and yet from that same blood is secreted a poison, which, if mingled with the blood of our system, will prove fatal to life in a very short time.

16. What other remarkable facts are noticeable in the action of the blood of the human system?

From the same atoms that enter into the formation of minerals and vegetables, the living blood is formed; by a different arrangement, in obedience to the laws of vitality in the animal system, from the matter composing this same living blood, the bone of the animal is formed; by a still different arrangement, the animal muscle is formed from the same blood; and by an arrangement still different from the others, from the matter of the same blood is formed the living animal nerve, which is the most remarkable, for its peculiar properties and powers, of any known material structure. All these are purely results of vital power, acting and accomplishing its ends as required by the body.

17. What is the vital force of the human body?

It is that power placed in the human body, at its birth, which will enable the body, under favorable circumstances, to live to a certain age. It is this which enables the body to rally and bring to bear its energies in throwing off disease.

It also battles against those influences that are liable to produce disease. It is spoken of in common-place language as *the constitution*. Of one it is said, "He will rally from that disease if his constitution is not broken." Of another, "He cannot rally, his constitution is gone;" meaning that either their *vital force* has so far been expended, or interfered with by violations of nature's laws, that it no longer has power to battle for the life of the body.

18. Can the original stock of this vital force be increased or diminished?

It cannot be restored when once expended, but it may be wasted, and life shortened proportionately. If the life force has been measurably wasted, by placing the person in the most favorable relations to life, his days may be protracted to a much greater extent than if he were left to follow out the ordinary habits of life. A realizing sense of these facts should certainly lead us to manifest the greatest care, lest we overtax our energies, waste our life force, and shorten our days.

19. How is the life of the body constantly maintained?

Chemical agents, and the physical laws of nature, are constantly exerting their influence on living bodies, causing an expenditure of vital power, and tending to the destruction of the vital constitution, and the decomposition of the organized matter. Therefore, life maintains a continual conflict with opposing forces; and hence it has been with truthfulness said, "Life is a forced state—a temporary victory over the causes which induce death."

20. What peculiarity is noticeable in the temperature of the human body?

The temperature of the human blood is, in a robust man, about ninety-eight degrees; and it hardly varies two degrees from this point, whether the temperature of the surrounding atmosphere be twenty degrees below zero or two hundred and sixty degrees above it. The animal body most completely resists the action of superficial heat and cold. The more vigorous the vital power is in animal bodies, the better are they enabled to sustain the extremes of heat and cold.

21. What can you say of the carbonic-acid gas thrown off by the human system in its action?

Carbonic-acid gas is thrown off in immense quantities by perspiration and respiration, and this, when received into the lungs, without a mixture of atmospheric air, is almost instantaneously destructive of animal life, but the vegetable economy, during the day, decomposes this gas, retains its carbon as vegetable nourishment, and sets free the oxygen, which is the peculiar principle of the atmosphere that supports animal respiration.

22. What is noticeable in the formation of the animal structure?

The most simple form of animalized matter composing the living body in the chyle, which is separated from the digested food in the alimentary canal, and enters the capillary tubes, by which it is conveyed to the blood vessels. This pearly-colored fluid, by chemical analysis, is almost wholly resolved into water. As it passes along the vitalizing tubes it becomes more and more albuminous and fibrinous. From the blood the vital economy of the body elaborates all the substances and forms of matter composing the animal body, constructing with marvelous skill and wisdom the blood vessels and the alimentary tube, with the assemblage of organs associated with it for the purpose of nutrition, and the outer walls of the body, with its limbs and organs of external relations. All the solid forms of the body, the bones, cartilages, ligaments, tendons, muscles, nerves, &c., are made from this fluid blood. They may all be reduced to three general kinds of substances: namely, the gelatinous, the fibrinous, and the albuminous, or, the cellular, the muscular, and the nervous tissue. The gelatinous substance, or cellular tissue, enters into the formation of the bones, cartilages, and tendons. It also forms sheaths for every muscle and for every cord of the nervous system. The fibrinous substance enters into the formation of the muscular tissue. The albuminous is the nervous tissue, which is the highest order of organized matter, and is endowed with the most peculiar and wonderful vital properties, and these properties are concerned in the functions of digestion, absorption, respiration, circulation, secretion, and organization, or the process of structure, and the production of animal heat.

23. What is the only element of positive motion in the human body?

With very limited exception, if any, the vital contractility of the muscular tissue is the only element of positive motion in the living animal body. Hence the muscular tissue is distributed wherever motion is required. The windpipe, stomach, intestines, heart, diaphragm, and several other internal organs are also supplied with this tissue.

24. What other arrangement is made in the human body for the security and protection of the organs?

The cavity of the body is divided by the muscular substance called the diaphragm, into two apartments. The upper one is called the thorax or chest, which extends from the neck to the breast-bone in front, and somewhat lower at the sides and back, and contains the lungs, heart, a portion of the large blood-vessels, and the esophagus, or food pipe. The lower division is called the abdominal cavity, and contains the liver, stomach, intestinal canal, pancreas, spleen, kidneys, &c. There is also a peculiar texture of the cellular tissue, called the serous membrane, which lines both cavities of the body, and is then extended and folded in such a manner as to envelop each organ separately, holding them in a measure in their proper place. This serous membrane in the upper portion of the body is called the pleura. It encloses each lung separately, and by two sheets, extending from the breast to the back, forms a double partition between the lungs. These two sheets are separated at the lower part of the chest to receive the heart. In a healthy state of the body the serous membrane has no animal sensibility. In fleshy people large quantities of fat are accumulated in many parts of this tissue. In a healthy action of all parts of the system, excess of fat never occurs, but waste and supply are equal. It must, from the considerations introduced in this chapter, be a matter of interest to all, to contemplate the subject of the following chapters, to learn what tends to waste our bodily structures, what habits of living will restore their proper action, and how we may thrive in mind and body.

## Chapter Two.

### THE HUMAN FRAME.

25. Of what is the human body composed?

Of solids in different degrees of density, and fluids that circulate through them.

26. What is the cubical size of the body, and what is the principal element of its composition?

The bulk of the body, upon an average, is equal to a cube of a little more than sixteen inches on a side. The principal element of the body is water. The amount of water equals a cube a little more than fourteen inches on a side, or nearly four-fifths of the body.

27. What are the solids of the body?

The solids of the body are bones, teeth, cartilages, ligaments, muscles, nerves, vessels, viscera, membranes, skin, hair, and nails.

28. What are the fluids of the body?

The fluids of the body are blood, chyle, lymph, saliva, gastric juice, pancreatic juice, synovia, mucus, and serum. Bile, sweat, and urine are excretions.

29. What does a chemical analysis of the body show?

It discovers to us that almost the entire bulk of the human body consists of Oxygen, Hydrogen, Nitrogen and Carbon. The bones and teeth are more than half phosphate of lime. The teeth also contain carbonate of lime.

30. Are there any other substances found in the body?

Yes; very small quantities of phosphorus,

sulphur, chlorine, iodine, bromine, potassium, magnesium, iron, aluminum, gold, lead, &c.

31. *What is the hardest solid in the body?*

With the exception of the enamel of the teeth, the bones are the hardest solid in the body.

32. *How are the bones constructed?*

The bony structure is a dense, sub-fibrous basis, filled with minute cells, and traversed in all directions by branching and connected canals called Haversian, which give room to blood vessels and nerves. These cells are irregular in form and size, and give off numerous branching tubes, which by communicating with each other constitute a very delicate network.

33. *What is found in the cavities and cells of bones?*

The internal cavities of long bones, and the canals and cells of others, are lined by a membrane, and filled with an oily substance called medulla, or marrow.

34. *Are the bones formed of what we eat?*

Yes; every part of the body is formed of and from what we eat, after the food has been changed into blood. As the blood circulates through the body, certain portions are secreted or separated from it to supply the several solids and fluids of the body.

35. *Is it then necessary that our food should contain the constituent elements of our bodies?*

It is. All substances containing these elements, however, are not proper food. Milk and eggs are supposed to contain nearly all the elements in the human body; but it does not follow from this that we should live wholly on milk and eggs, nor that we should eat lime, or drink lime water, because there is lime in our bones. Fruits, grains and vegetables, contain every element composing the human body, and that, too, in a state easy of being appropriated by our system to build up the structures of the body. But more of this under the head of digestion.

36. What is the strength of human bones?

Human bones, when used as levers, are twenty-two times as strong as sandstone, three and one-half times as strong as lead, nearly two and three-fourths times as strong as elm and ash, and twice as strong as box, yew, and oak timber.

37. Does the quality of the food we eat affect the strength and soundness of the bones?

It does. If our food is not sufficiently nutritious, or is of too poor a quality, our bones will be liable to be soft and diseased. This is the most effective cause of the rickets. As the bones become softened, by the strength of the muscles the body is drawn into unsightly deformity.

38. What other means injure the bones?

Too little exercise in the open air, working in mines, working or living in damp, or poorly-lighted places, sleeping in close rooms, or rooms where the air is stagnant or impure, or keeping our bodies, while laboring, constantly bent, or in any posture which prevents the free circulation of the blood, and the natural action of the vital organs; all these injure the strength and health of the bones. Children, especially, should not be confined in any unnatural position, but be allowed to move freely in whatever direction nature may demand.

39. How often is it supposed our bones undergo a change?

In from one to ten years it is supposed that the entire body, including the bones, undergoes a change. This change is caused by the minute particles that form the body undergoing a state of decay and reproduction. This change, however, is so gradual—particles passing off and others taking their place—that the body, to a great extent, retains its identity through life.

40. Where are the bones of the human body placed?

They constitute the frame on which the body is built. They give form and strength to the body, support its various parts, and prevent it from sinking by its own weight; they serve as levers for muscles to act upon, and to defend the brain, heart, lungs, and other vital parts, from external injury, and occupy the same position in the body that the frame does in a building. The muscles, nerves, flesh and skin, are placed upon the bones as a carpenter puts boards on the frame to build the house.

41. How many bones are there in the human body?

The number is variously estimated by different anatomists from 240 to a much larger number. The best authorities, however, give 246 distinct pieces in the body of a grown person.

42. How many kinds of bones are there in the human body?

Three: long, flat, and irregular. The long appertain to the limbs, the arms, legs, fingers and toes; the flat inclose cavities, as the brain and pelvis; the irregular are formed mostly about the base of the skull, face, trunk, wrist, and instep. All these forms of the bones are requisite for the situations they occupy, and the respective functions they fulfill.

43. What is the only bone in the body which is completely *ossified*, or hardened at birth?

It is that bone which is called the *petrous*, which contains the organs of hearing. The bones do not become solid till the twelfth year of life.

44. Are the bones of the young liable to become otherwise injured?

Yes; many persons in making their little children sit alone at too early an age, produce in them a crooked spine. In allowing them to stand or walk before the bones of the legs are sufficiently toughened, their legs become crooked, either bandy-legged or knock-kneed, for life. It is for this reason important that great care should be taken while the bones are soft, that they be not misshaped. Children should not be urged to walk. They will try to walk themselves when their bones are sufficiently toughened to walk. The flat-head Indians of North America tie hard pieces of board on the back and front side of the heads of their children till the skull hardens in this shape, which causes the head to have its flattened appearance, which it retains for life.

45. Are the bones of old people as strong as those in middle age?

No; in most cases the bones of the aged are dry and brittle, hence are more easily broken by a fall than those of younger persons. When the bones of the aged are broken, the process of knitting the bone together, as it is called, goes on, if at all, very slowly. For this reason, it requires two or three times as long a period for one to get about with a broken limb at seventy years of age, as for one at twenty-five or thirty.

*Figure I.*

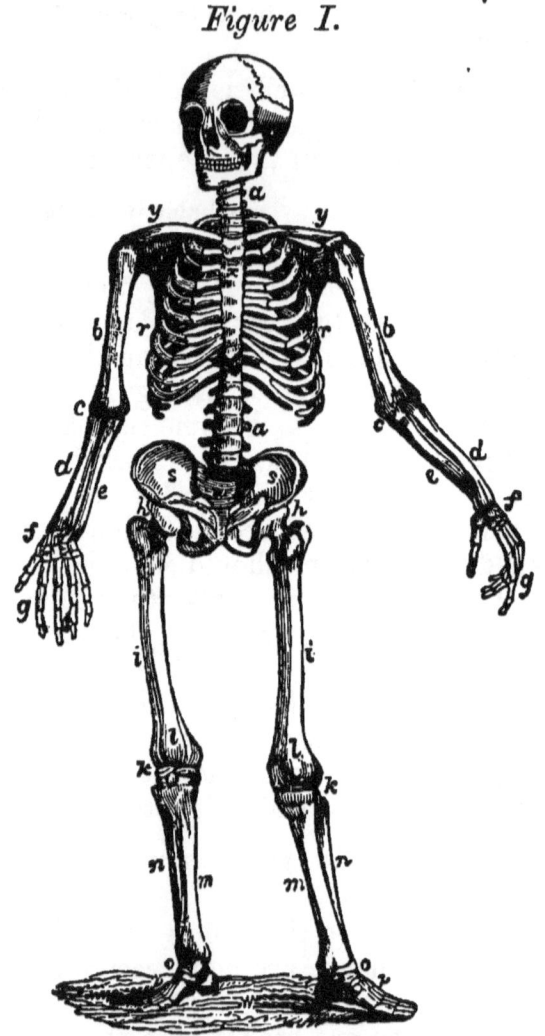

HUMAN SKELETON.

In *Fig. I*, we have a human skeleton. *a a* is the back bone, at the upper part of which is the axis on which rests the *atlas*, as it is called; *b*, humerus, long arm bone; *c*, joint of elbow; *d*, *e*, ulna and radius, the two bones of the fore-arm; *f*, the carpus, or bones of the wrist; *g*, metacarpus and phalanges, bones of the hands and fingers; *h*, the joint of the hip; *i*, the femur, or thigh bone, the longest bone in the body; *l*, the lower end of the femur or thigh bone, which is enlarged; *k*, the patella, or knee-pan; *m*, the tibia; *n*, the fibula; the two bones of the leg; *o*, the tarsus, or bones of the heel and instep: *p*, metatarsus, bones of the foot and toes; *r*, the thorax, or bones of of the chest, ribs, &c.; *s s* and *w*, the pelvis; *w*, the sacrum, a wedge-shaped bone at the lower end of the back bone; *x*, the sternum, or breast bone; *y*, the clavicle, or collar bone, which extends across the upper part of the chest, from the upper end of the sternum to the shoulder blade.

46. What is the average weight of a human skeleton?

About one-tenth the weight of the whole body.

47. What are the bones of the human skeleton, and how many are there of each?

| | |
|---|---:|
| Bones of the skull, | 8 |
| Ear, | 6 |
| Face, | 14 |
| Teeth, | 32 |
| Back, vertebral column, | 24 |
| Ribs, twelve pairs, | 24 |
| Tongue, os hyoides, | 1 |
| Upper extremities, arm, wrist, & fingers, | 64 |
| Breast bone, sternum, | 1 |
| Pelvis, hip, sacrum, and coccyx, | 4 |
| Lower extremities, leg, instep, and toes, | 60 |
| Sesamoid—knee pan, and bones in tendons, | 8 |
| Total, | 246 |

48. What are the bones of the back called?

The vertebral column.

49. How many pieces are there in the vertebral column, *a a, Fig. I*?

Thirty-three in the young person, but in advanced life the nine lower pieces unite into two.

50. What is each of these pieces called?

Vertebra. Twenty-four of these are called the true vertebræ, the rest are called the false vertebræ, which unite to form the sacrum and coccyx; these last are also concerned with the hip bones in the formation of the pelvis or basin at the bottom of the trunk, and constitute the base on which the vertebral column rests. Of the true vertebræ, seven belong to the neck, twelve to the back, and five to the loins; and are accordingly distinguished by the terms cervical, dorsal, and lumbar vertebræ, from the Latin *cervix*, neck; *dorsum*, back; and *lumbar*, loins.

51. What is the first cervical vertebra, which supports the head, called?

Because it supports the head, it is called the atlas. This is said to be thus named from the tradition that a giant by the name of Atlas supported the earth on his shoulders.

52. What is the atlas?

It is the ring of bones which allow the head to move sidewise as well as backward and forward to some extent on the second cervical, which is called the axis.

53. Are the bones of the vertebræ solid?

No; there is a large cavity the whole length of the spinal column for the spinal marrow, and two smaller cavities each side of the spinal marrow, extending down along the back side of the vertebral column, for the spinal cord.

54. What do the bones of the spinal column resemble?

They have some resemblance to so many rings piled upon one another. This is not an exact resemblance however, for they have several projections from the arch behind; one running directly back, which is called the spine. Two running obliquely backward, with which the ribs form one of their joints of attachment. The vertebræ are therefore so constructed, that when arranged in their proper order, they form both a column of support to the body, and a canal for the spinal marrow. Between all of these bones is interposed an elastic, fibrous cartilage, which, with the surrounding ligaments, unites and binds them to each other in such a manner as to give the column considerable flexibility and elasticity, and at the same time secure to it all the supporting power of a solid bone. Thus it forms a strong upright column, which gives erectness, dignity, and grace to the human body.

55. How do many persons injure the shape of the spinal column?

By wrong positions in sitting, standing, or lying down. By sitting considerable of the time, as many do, in rocking chairs, or while writing, bent forward, or with one shoulder higher than the other. By these ill-habits, this column becomes bent too far forward, or crooked sidewise, causing either round shoulders, or a dropping of one shoulder lower than the other. Some lie on two or three pillows, so that when they habitually lie upon the side they are in danger of causing this same curvature of the spine. In sitting, you should sit back against the back of the chair, with head erect, shoulders back, and the whole vertebral column to the shoulders resting against the back of the chair. In lying down, whether on the back or side, lie with the body, arms, and limbs straight, and the head elevated not more than four inches. You should habituate yourself to sleeping on either side. Frequently changing from side to side is also beneficial. Never sleep lying upon your face.

56. Into how many parts is the skull divided?

Four; superior, lateral, inferior, and anterior. The superior is the front and upper portion, or that containing the intellectual brain. The lateral, is the sides. The inferior, the base of the head. The anterior, the face.

57. What are the cavities of the skull called in which the eyes are placed?

The orbits of the eyes. These are hollow cones for the lodgment of the eyeballs with their muscles, vessels, nerves and glands.

58. How many bones are there in the head?

Sixty-one, above the upper joint of the neck, including the teeth.

59. How many bones are there that give shape to the skull?

Eight, and these are united together at their edges. These edges which lap into each other somewhat resemble saw teeth. These ragged pieces of bone uniting the parts of the skull together (see *Fig. 1*) are called *sutures*, from the Latin *sutura*, to sew, because they look very much like the seam made by sewing two pieces of cloth with the "over-and-over" style of stitch.

# The Human Frame.

60. How many bones are there in the face?

Fourteen, besides the teeth.

61. Are there any bones in the ear?

Yes; there are three bones in each ear, and these assist in conveying sound to the brain.

62. Are there any bones in the tongue?

Yes; there is one bone at the root of the tongue called os hyoides. It is used to support the tongue and upper part of the larynx, or windpipe.

*Figure II.*

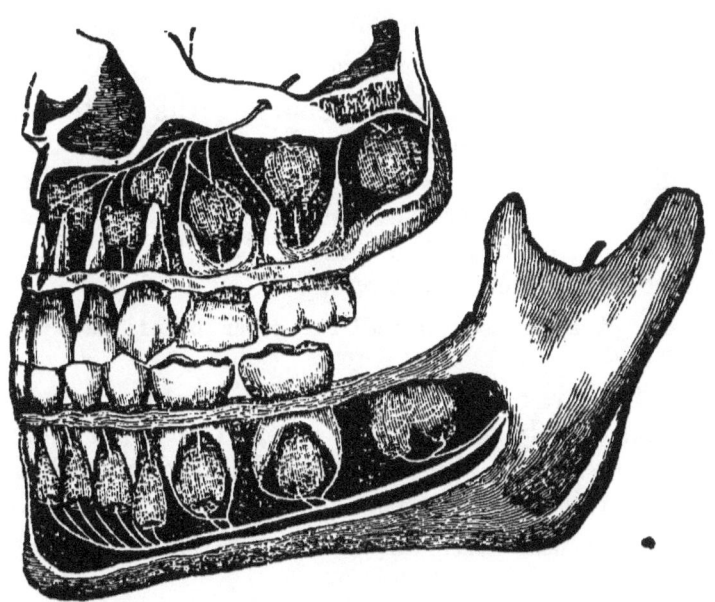

THE TEETH.

63. What does *Fig. II* illustrate?

It illustrates the formation of the teeth, their nerves, &c. The infant set are in the jaws, while the outlines of the second set are also observable.

Keep your eye on this figure while answering the following questions.

64. How many teeth has the human being?

The permanent teeth are thirty-two in number, sixteen in each jaw.

65. How many parts are there to the tooth?

There are three parts; called the crown, the neck, and the root. The crown is that part which is seen above the jaw, the neck is that portion clasped by the upper rim of the socket, and the root is that part within the gum and socket, which is fastened to the jawbone.

66. Of what are the teeth composed?

Of a firm crust, called enamel; the tooth bone proper, called the ivory, and a cortical substance, called cementum. The enamel covers the exposed surface of the crown, and the cementum forms a thin coating over the root of the tooth. These become thinner in old age.

67. Are the teeth, like all our other bones, made from our food?

They are; and like the other bones, are composed largely of lime, but unlike our other bones, are exposed to the immediate action of the air and foreign substances.

68. In what other respects do the teeth differ from the other bones?

They are composed of a much harder material. The ivory of the tooth is much harder than bone, and the enamel is still harder.

69. What is the use of this enamel?

It gives the teeth strength, as well as hardness, for biting, chewing, and grinding the food; it also prevents injury from these operations, and from the action of acids on the bone of the teeth; it also adds much to their beauty.

70. If this enamel is broken, does it form on the tooth again?

When the enamel is once destroyed it is seldom, if ever, restored again. When it is once broken the teeth are liable to rapid decay. It is important to use our teeth carefully in this respect, and not use them to crack nuts, or bite very hard substances, lest in after time we be deprived of their more important use in grinding our food.

71. Are the teeth supplied, like our other bones, with blood vessels and nerves?

They are; and as most people have occasion to know, these nerves are endowed with life, and also an exquisite sensibility, which is the more apparent when the teeth become decayed.

72. How many teeth has a young child?

Twenty; ten in the upper, and ten in the lower jaw.

73. What care is necessary in relation to the teeth of children?

When a child is from five to seven years of age, the teeth loosen, when they should be immediately removed; otherwise they will prevent the proper formation and regularity of the new and permanent teeth, which are growing under them. Some persons permit their children to eat candies and sweetmeats with their first set of teeth, and manifest but little care for them; but it should be borne in mind that the same nerves and blood-vessels that are connected with the first set of teeth, are at the same time communicating with the embryo forms of the second set, which are forming beneath them. See *Fig. II.* The nature of the second set in a great measure depends on the nature and care of the first set. Disease of the child's teeth then may cause them many ills in after life.

74. Why is it necessary for the human mouth to be furnished with two sets of teeth?

The gradual growth of the body, renders it necessary that our little jaws should be furnished with a set of teeth in childhood, which are too small to fill up our jaws when our system is fully developed, and too small to answer the purposes of mastication through life; and hence the all-wise Creator has established a law in our system by which the small teeth of our childhood are removed, and their places supplied with a larger, permanent set, which are better fitted for mastication.

75. What is the difference between the first and second sets of teeth?

The first teeth, or those of a child, only pass through the gum socket which is fastened to the jaw, while the second set grow out of the jaw itself, between the roots of the first set of teeth; so if the first are not removed, the second set must force their way inward or outward between them.

76. How many different kinds of teeth have we?

Three; four cutters, or front teeth, in each jaw; two pointers, or eye teeth, in each jaw; and ten grinders, or back, double teeth, in each jaw; half on each side of the face.

77. What is the most important use of the teeth?

Their leading and most important use is to cut and chew, or grind the food so finely that it may be mixed with the saliva, or the moisture of the mouth, before passing into the stomach.

78. If we had no teeth, would we have the pleasure in eating we now enjoy?

No; for then our food would need to be mostly liquid or semi-fluid.

79. Are the teeth otherwise useful?

Yes; they assist the voice in talking, reading aloud, and singing. If a person loses two or three front teeth, he talks, reads, and sings, in a hissing, disagreeable manner. The loss of teeth prevents a person from giving the correct sounds of many letters, and from articulating distinctly.

80. Should we not do everything in our power to preserve our teeth?

Yes; we should never pick nor scratch them with pins or pocket knives; for these break the enamel. Quill or wooden tooth picks may be useful in removing any particles of food that may not be readily reached by the brush, but metallic tooth picks should never be used.

81. In what other ways are the teeth injured?

By taking into the mouth food or drink which is either too hot or too cold, by smoking or chewing tobacco, by using acid drinks or fruits which set the teeth on edge. Hot substances taken into the mouth serve more directly and powerfully to destroy the teeth than any other cause which acts immediately upon them.

82. Why are the teeth of Europeans generally better than those of Americans?

The principal reason is, their food is more simple, and their habits more temperate and uniform, than those of Americans.

83. How can we care for the teeth?

The teeth should be cleansed with a brush or a soft piece of flannel, and tepid water, after every meal, but more especially before retiring to rest, and again after rising in the morning. Some refined soap may be occasionally used, to remove any corroding substance that may exist around or between the teeth. The mouth should be rinsed after its use. Soft water is always best for the teeth. If the teeth are closely set together, drawing a thread between them occasionally will be of great benefit.

84. What is the cause of the pain called "tooth-ache?"

When a tooth is so decayed that its inflamed nerve is exposed to the air, it causes tooth-ache. Sometimes food crowded against the bare nerves in eating, produces the same effect.

85. What should we do with decayed teeth?

If any of our teeth have begun to decay, a dentist should be consulted as soon as possible, and the cavities filled with gold. Natural teeth, if partly filled with gold, are always better than artificial teeth. When teeth are past filling, they should be immediately removed, otherwise they will cause decay in adjoining teeth, give rise to neuralgic pains, or cause maxillary abscess, which is known by a severe and obstinate pain in the face, just below the eye, near the nose. Sometimes this disease causes discharges of offensive matter from the nose, it also produces bad breath, and affects the general health. It may be years in forming, and be mistaken for common tooth-ache.

86. What is another great leading cause of the premature decay of teeth?

Their disuse. The more the teeth are regularly and properly used for the purposes for which they were intended, that of masticating and preparing the food for the stomach, the more healthy they will be, and the less liable to decay. Experience shows that the teeth decay the most rapidly between the ages of fifteen and thirty. So that youth need to give the most special attention to their teeth.

87. Who have generally the best teeth?

Those who have the best health. Therefore to assist in preserving the teeth, the stomach and lungs should be kept in as healthy a condition as possible. The proposition we think is correct, that diseases of the nervous system affect the teeth, and also diseased or decaying teeth have a powerful effect upon the general health. The loss of the teeth cripples the natural action of the system—lessens the action of the salivary glands, and to some extent shortens life.

88. What are the bones of the chest?

The sternum, or breast-bone in front, and the twelve pairs of ribs on the side, and these constitute the thorax. See *Fig. I.*

89. What is the sternum, or breast-bone?

It is that bone which lies directly in the central line of the fore part of the chest; its upper end lies within a few inches of the vertebral column, while its inferior extremity projects considerably forward. It is about eight inches in length, and one and a half inches in width.

90. What is the form of the ribs?

They grow out of the spine, or back bone, on the back side, forming a hoop by meeting and being fastened to the breast bone in front.

91. *Do all these ribs grow directly to the breast bone?*

The first or upper seven pairs, grow directly to the sternum. The five lower pairs are called false ribs, and are connected with each other in front by cartilages, or a substance somewhat like bone, but more pliable and spring-like.

92. *Are all the ribs of one size and shape?*

No; they increase in length from the first to the eighth, and then diminish in length to the twelfth. In breadth they diminish from the first to the last, except the two lower ones. The first is horizontal and all the rest oblique. The two lower ribs are called floating ribs.

93. *Of what use are the ribs?*

They are the frame-work of that part of the human trunk termed the chest, in which the lungs and heart are deposited for safe keeping.

94. *Is it important to care for this frame-work?*

It is. If we wear our clothing too tight, we diminish the size of the chest, crowd the lungs, heart, and other organs, and hinder their healthy action. By sitting, or standing in a stooping posture, the lower end of the sternum is crowded upon the stomach, which injures and weakens it. Men or women who wear tight clothing over the lower ribs must injure their health.

95. *What do the bones of the upper extremities comprise?*

They are the clavicle, or collar bone; the scapula, or shoulder blade, a large, flat, triangular bone back of the shoulder; the humerus, or arm bone; the ulna and radius, or bones of the forearm; these two move at the elbow similar to a door hinge. The word *ulna* is Latin, and means an ell. This bone is so named because in early times it was used as the measure of an ell; the carpus, or the eight bones in two rows across the wrist; the metacarpus, or five bones of the palm of the hand; the phalanges, or fourteen bones of the fingers.

*Figure III.*

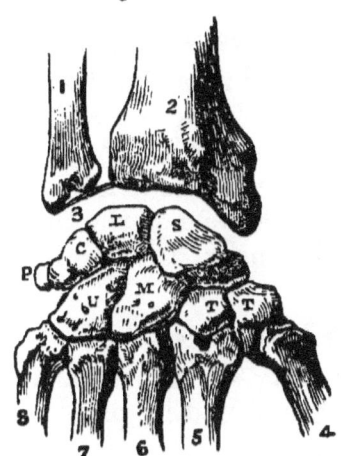

THE CARPUS.

96. What does *Fig. III* represent?

The carpus. 1, lower end of the radius; 2, lower end of the ulna; 3, connecting cartilage; *S L C P*, first row of bones in the wrist; *T T M M*, second row of bones; 4, 5, 6, 7, and 8, the five bones of the palm of the hand, to the end of which the phalanges, or finger bones are attached.

97. What is the pelvis?

The pelvis is that part of the human body which is fixed in relation to the rest, and about which all the rest move; beneath it the lower extremities walk, and above it the back bone is movable. It is a collection of bones at the lower end of, and attached to, the vertebral column. The upper part of it serves as a support to the abdominal organs. To the lower sides of it the thigh bones are attached. See *s s* and *u*, *Fig. I.*

98. What are the bones of the lower extremities?

The femur, or thigh bone, which is the longest bone in the body; the patella, or knee-pan, which is heart-shaped in figure, and is connected with the lower portion of the thigh bone; tibia and fibula, or the bones of the leg; the tarsus, or seven bones of the ankle; the metatarsus, or the five bones of the foot; the phalanges, or fourteen bones of the toes. The bones of the tarsus and the metatarsus correspond, somewhat, to those of the carpus and metacarpus of *Fig. III.*

99. Are there any other bones in the human body?

Yes; there are what are called sesamoid bones. These are formed in tendons, and are a sort of pulley on which the tendons of the body play. They are also placed over joints. The knee-pan is one of them. There is one of them on the joint of the thumb, and of the little finger, and in other parts.

100. Are the bones of the body mostly in pairs?

All the two hundred and forty-six bones of the body, except thirty-four, are found in pairs, or one on each side of the body.

101. What is the connection between any two bones called?

A joint or articulation. It is by means of these joints that the various motions of the bones are easily made.

102. How many joints are there in the human body?

Over two hundred, all perfectly adapted to their various positions and work.

103. How many motions can be made by the joints?

The motions of the joints are four. The first is where the bones slip over one another, which is the case in all joints. The second is angular movement, as up and down, right and left, or such motions as you can make with your fore finger sidewise, and by bending and straightening it out. The third motion is circular. It consists in moving the bone around in a circle with the joint as a center, such as you make in whirling your arm around the shoulder joint. The fourth motion is called rotation, such as the moving of the atlas on the axis in the neck.

104. What is the construction of a movable joint?

The opposing surfaces are coated by an elastic substance called cartilage, this is lubricated—oiled—by a fluid called synovia, which is secreted in an enclosed membrane or bag, called synovial.

105. How are the bones held firmly together?

By bands of glistening fibers, called *ligaments*. These are mostly short, and attached only to the enlarged extremities of the bones. In such joints as that of the shoulder and hip, there is between the ends of the bones what is called the *round ligament*. It is a bundle of fibres in the form of a cord. It is used to keep the head of the bone from slipping out of the socket, and at the same time it allows the most perfect freedom of motion.

106. How are the joints kept in their places, or the bones from getting out of joint when they are moved?

There are ligaments formed about all the joints, sometimes constituting bands of various breadths and thicknesses, and sometimes layers of these bands are extended around the joints. These serve the same purpose in the human frame as the pins in a frame building. Instead, however, of clumsy joints being made and pinned together, a few tough fibres and membranes, secure at once, in a most perfect manner, every portion of the frame, and provide at the same time means for its lubrication. Some of the ligaments are situated in the joint, like a central cord or pivot, and some surround it like a hood.

107. Where are the ligaments principally found?

The ligaments bind the lower jaw to the temporal bones, the head to the neck, extend the whole length of the back bone in powerful bands, both on the outer surface, and within the spinal canal, and from one spinal process to another; and bind the ribs to the vertebræ, and to the spinal projection behind, and to the breast bone in front, and this to the collar bone, and this to the first rib and shoulder blade, and this last to the bone of the upper arm at the shoulder joint, and this to the two bones of the fore-arm at the elbow joint, and these to the bones of the wrist, and these to each other, and to those of the hand, and these last to each other and those of the fingers and thumb. In the same manner they bind the bones of the pelvis together, and the hip bones to the thigh bone, and this to the two bones of the leg and knee-pan, and so on to the ankle, and foot, and toes, as in the upper extremities.

108. What can you say in general of the cartilages and ligaments?

They unite, and bind the whole bony system together in a powerful manner, so as to possess in a wonderful degree mobility and firmness. The ligaments and cartilages are in health destitute of animal sensibility. They are soft and yielding in early life, and become more dry, rigid, and inflexible in old age.

109. What other useful contrivance is connected with the joints?

In all the movable joints the articulating surfaces of the bones are covered with dense and highly-polished cartilages, by which means the joints are enabled to act with great ease and little friction, and at the same time it constitutes a cushion which breaks in a great measure the sudden jar which otherwise would be felt in the head, when walking or suddenly moving the limbs. It protects the brain in the same manner that the springs of a carriage prevent a sudden jolt. Cartilage is also employed separately from the bones in forming some of the cavities, &c., as the larynx, wind-pipe, part of the nose, &c.

110. How many kinds of joints are there?

Three: fixed, or such as the joints of the skull and upper jaw, teeth and vomer;* movable, such as the shoulder, hip, elbow, wrist, knee, ankle, carpus, and tarsus; intermediate, or such joints as those in the vertebral column.

111. Why are the bones of the body mostly cylindrical, or hollow?

To secure great strength with as little material as possible. Were the bones of the human skeleton made solid, it would be so heavy and cumbersome that it would require a larger amount of muscles, making the body unwieldy, and thus depriving it of its rapidity and ease of motion.

112. Is there any other peculiarity in the bones not already mentioned?

The bones are closely covered with a very firm,

---

*The gristly subtance between the nostrils is attached to what is called the vomer.

whitish-yellow membrane, very smooth, this is called the periosteum. This membrane encloses the vessels which carry nutriment into the bones. It is to this periosteum that the ligaments and tendons are attached, as they cannot fasten to the bone itself. In fever sores and felons the disease begins in most cases in the periosteum of the bone, or the tendons, but if not checked it soon affects the bone itself. Cool baths, or a cool wet bandage upon the affected part during its fevered condition is good. Felons may many times be driven away by keeping the affected part immersed a long time in hot water. But much time and suffering can be saved by cutting a free gash through the felon into the bone through the periosteum. Cut a gash an inch long, at least.

113. Are there other diseases, or difficulties with the joints, or bones, that we should guard against?

Yes; strains of joints, and dislocations, many times caused by wrestling. This is a dangerous exercise, and should give place to milder sport. Inflammation in sprains should be allayed with the cool wet bandage, covered, of course, by dry flannel. Violent jerking of the arms in gymnastics should be avoided, as it is liable to cause synovitis, or disease of the joints.

## Chapter Three.

### THE MUSCULAR SYSTEM.

*Figure IV.*

MUSCULAR SYSTEM.

114. What are the muscles?

The moving organs of the body. Their grand peculiarity is their power of contraction, which is as wonderful as anything in nature. This is the element of all voluntary motion, and most, if not all, positively involuntary motion, in the living body. All the great motions of the body are caused by the movement of some of the bones which constitute the frame-work of the system; but these, independently of themselves, have not the power of motion, and only change their position through the action of other organs that are attached to them, which by contracting, draw the bones after them. In some of the slight movements, as the winking of the eye, no bones are displaced. The organs which perform this remarkable work are called *Muscles*. The red color of the muscles is owing to the presence of the numerous blood vessels which they contain.

115. Of what are the muscles composed?

They are composed of parallel fibers, of a deep red color, constituting lean flesh. Any person can examine a piece of boiled beef, or the leg of a fowl, and see the structure of the fibres and tendons of a muscle.

116. Of what shape is a muscular fiber?

It is long, round, and fine, like a thread. Any piece of lean meat can be easily divided in one direction into stringy fibers, which, by the use of appropriate instruments, can be subdivided, till fibrils are reached, not so large as hairs, composed of a sheath enclosing minute particles, shaped like beads, placed end to end. Some might suppose lean meat had no regular shape, but the above proves that it has. The lean portions of our bodies are arranged in perfect order, as may be seen by observing *Fig. IV*, at the head of this chapter.

117. How are the fibrils that constitute muscles held together?

By a delicate web or sheath, which is perforated with minute tissues or cavities, and these become so compact together at the ends of the muscles as to form glistening fibers and cords, called tendons, or sinews, by which the muscles are attached to the periosteum, or surface of the bones. A few of the muscles resemble in structure a ribbon; others a cord; others are thin and expanded, so that they resemble a membrane. The muscles present various modifications in the arrangement of fibres, as relates to their tendinous structure.

118. How many muscles are there in the human body?

It is supposed that there are not less than four hundred and seventy distinct muscles of voluntary motion in the human body; about twice as many muscles as there are bones. They are nearly all arranged in pairs, each side of the body having the same kind. These are so arranged and adjusted, as to position and connection, that by the contractions of the different pairs, or individual muscles, all the voluntary motions of the lower limbs are performed. The function of respiration —which to a certain extent, is both voluntary and involuntary—also employs some of these muscles.

119. How much of the body is muscle?

The greater portion of the bulk of the body is composed of muscular tissue. It is the muscle that gives the body its plump appearance. These muscles not only serve as a means of moving the body, but, in the limbs, they invest and protect the bones, and some of the joints. In the trunk they are spread out to enclose cavities, and form a defensive wall, capable of yielding to external pressure and again returning to its original position.

120. How are muscles moved?

By contracting, or shortening. When a piece of India rubber has been stretched and you let go of it, it contracts. These cords have power to lengthen and shorten somewhat like rubber. Some muscles in contracting pull at one end, and some at both their ends.

121. How are all the actions and motions of the various organs of the body produced?

By the alternate contracting (shortening) and expansion (lengthening) of these muscular fibers. These muscles are so arranged as to act as antagonists to each other, some *di*splacing a part, and some *re*placing it; and therefore they are termed the flexor and extensor muscles. The flexor muscles are considered to be generally more powerful than the extensor, and hence, when the WILL ceases to act, as in sound sleep and death, the body and limbs are partially fixed or bent.

122. Of what are the muscles composed?

Of bundles of fibers enclosed in a sheath; each fiber is composed of smaller bundles, and each bundle of single fibers called *ultimate fibers*. By a microscope it is seen that these ultimate fibers are composed of finer fibers called *fibrils*.

123. What is the appearance of the end of one of these fibers through a microscope?

It presents to us an appearance similar to that of the end of a compact bundle of very fine straws.

124. How many kinds of muscular fiber do anatomists distinguish?

Two: those of voluntary or animal life, those under control of the will; and those of involuntary or organic life, such as are used for breathing and digestion. Muscular fibers of animal life are composed of these bundles of fibrils, while the muscular fibers of organic life are flat, and are held together by fibers which are composed of a dense form of the same tissue. The muscles of animal life are developed on the external part of the body and are mostly attached to the bones, and they comprehend all of the muscles of the limbs and trunk, while the muscles of organic life are formed from the internal or mucous layer, and are situated in the hollow organs composing the respiratory, digestive and circulatory apparatuses.

## The Muscular System.

125. When the muscular tissue is once destroyed, is it ever restored again?

Never; but when the muscles are wounded, with or without the loss of their substance, the breach is healed, and the parts united, by what is called *areolar* tissue, which is wholly insensible to the action of stimulants.

*Figure V.*

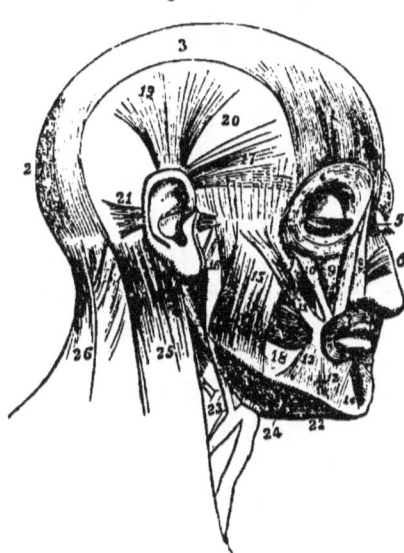

MUSCLES OF THE HEAD AND FACE.

126. What does *Fig.* V illustrate?

This figure illustrates the muscles of the head and face. 1 is the muscle moving the eyelids; 2, muscle used in drawing the top of the head backward; 3, point of attachment of muscles 1 and 2; 4, muscles used in opening and closing the eye; 5, muscle that draws down the corner of the eyelid; 6, muscle used to expand the nostrils; 7, muscle surrounding the mouth, used in closing the lips; 8, muscle used to elevate the upper lip and expand the nose; 9, muscle used to elevate the upper lip; 10, 11, muscles that pull the angle of the mouth upward and outward, and are used in laughing; 12, the muscle which pulls the lower lip downward and outward; 13, muscle which pulls the angle of the mouth downward and outward, is used in making expressions of grief;

14, muscle that raises and protrudes the chin; 15, 16, 18, muscles used to move the jaw in chewing; 17, 19, 20, muscles which in lower animals move the ear; they have but little motion however in the human body; 20, the covering of a muscle; 22, 23, 24, muscles which, with their attachments to the neck, are used to move the lower jaw downward; 25, 26, muscles used in the movements of the head and shoulders.

127. How many groups of muscles are there in the head and face?

There are eight groups of muscles, namely: 1 for moving the eyebrows; 3 for moving the eyelids; 7 for moving the eyeballs; 3 for moving the nose; 7 for moving the lips; 3 for moving the chin; 5 to assist the lower jaw in the motions necessary in masticating food. One of these muscles passes over the temple, and is called *temporal*, from *tempus*, time, because here the hair begins to turn gray; 3 muscles of the ear. The muscles used for chewing food are attached to the lower jaw, near the joint; if their position was near the front part of the bone they would not contract sufficiently to bring the jaws together.

128. How many groups of muscles are there in the neck?

There are eight: The first has 2 muscles, used to bow the head forward; the second has 8 muscles, 4 depressors and 4 elevators. The depressors pull down the os hyoides, or bone of the tongue. The 4 elevators raise the os hyoides when the jaw is closed; the third has 5 muscles, and these aid the tongue in all its movements; the fourth has 5 muscles, which are used in swallowing food; the fifth has 3 muscles, used in all the motions of the palate; the sixth has 5 muscles, passing from the sides of the head down on to the breast, used to steady the head, and to lift the ribs as we draw the head backward in inhaling a long breath; the seventh consists of each of those muscles used in varying the tones of the human voice; the eighth has 8 muscles, those of the larynx—Adam's apple; also used in producing articulate sounds. The muscles of the above eight groups, then, are those adapted to move the head and neck on the spinal column, to raise the shoulders, to control the motions of the mouth and throat, and to produce the sounds of the voice.

129. How many layers of external muscles are there in the back?

There are six layers, composed of at least thirty pairs of muscles. They give a firm attachment for muscles to move the extremities, and keep the trunk in an upright position. The first layer has 2 muscles and the second has 3, and these two layers give all the different motions to the shoulders; the third layer has 3, some of which are used in raising and depressing the back portion of the ribs in breathing, and some are used to give the spinal column its slight motion from side to side; the fourth layer has 7 muscles, holding the vertebral column erect, and assisting in steadying the head. When one of the muscles of this character act on one side alone they produce the rotation of the atlas on the axis; the fifth layer has 7 muscles. A portion of these contribute to the support of the back in an erect position. The others produce the rotary and other motions of the atlas on the axis; the sixth layer has 5 muscles. A portion of these raise the back part of the ribs in the inspiration of breath, the others help in supporting the body and holding the bones in position.

130. How many classes of muscles belong to the thorax or chest?

There are three classes of muscles belonging to the chest, and these belong also to the upper extremities. The first class is 11 internal muscles; the second class is 11 external muscles. These muscular, cordy fibers fill the spaces between the ribs outside and within the front of the chest, and they cross each other; the third class is situated within the chest, and connects the breast bone with the ends of this first and second class of muscles, and also connects them with the cartilages of the second, third, fourth, fifth, and sixth, ribs. The lower fibers of this muscle connect also with the internal muscles of the abdomen. These three classes of muscles raise and depress the ribs and draw down the cartilages of the ribs, and thus assist in breathing.

131. How many muscles regulate the movements of the abdomen?

Nine, and the ninth of these is called the diaphragm. It is a partition across the body just below the lungs, with an opening near the center. It separates the thorax from the abdomen. Its point of attachment to the body is just below the ribs. When the diaphragm is relaxed, it presents the appearance of an inverted and irregularly-shaped cup. When in a state of contraction its surface is nearly a plane. This muscle enlarges the chest by depressing its lower surface, as in the case of holding a full breath in the lungs. The nine muscles of the abdomen assist in drawing down the ribs, in drawing in the small of the

THE MUSCULAR SYSTEM. 51

back, or in turning the abdomen sidewise. They also diminish by their action the size of the abdomen, and thus assist in breathing.

*Figure VI.*

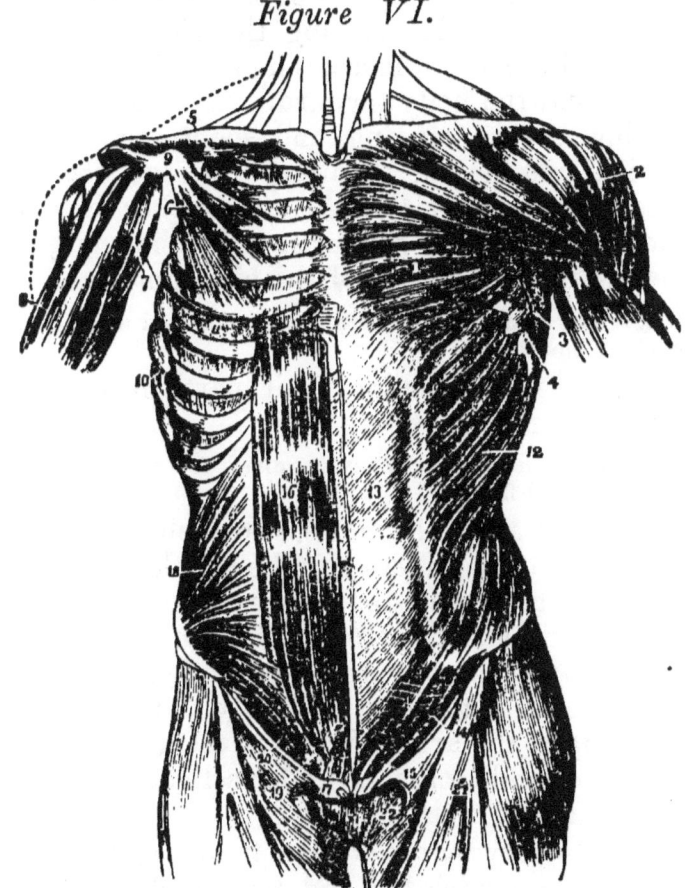

MUSCLES OF THE TRUNK.

132. What does *Fig. VI* represent?

In *Fig. VI* are seen the front muscles of the trunk. 1, 2, 3, are muscles used in moving the shoulder; 4, a muscle raising the ribs by the movement of the shoulder. It is this muscle which draws the shoulder forward in a case of diseased lungs; 5, 6, muscles used in raising the shoulder, or in producing its rotary movements; 7, 8, 9, muscles used in movements of the arm and shoulder; 10, 11, muscles moving the ribs in breathing; 12, 13, 16, 18, muscles moving the abdomen, and to bend the body forward. They are used in laughing, crying, singing, and shouting; 14, the principal muscle for throwing the leg forward, called *Poupart's ligament.* The other numbers indicate the muscles supporting the abdomen, and assisting in the motions of the hip joint.

133. How many muscles are there connected with the outlets of the bowels?

There are 8, all acting their part in carrying waste matter from the system.

134. How many muscles are required for the movements of the arms and fingers?

There are about 46 muscles to each arm. And in order that the fingers may be slender, and easily moved, the muscle that moves them is placed in the fore arm, and the tendons for the movement of the hands and fingers are made as slender as possible. Some of these extend only to the wrist to bend it upon the arm. Some of them pass to the very finger ends, passing under a ligament like a bracelet, at the wrist.

135. How many muscles are there in each leg and foot?

About 52 in each leg and foot, and these are used to make all the different motions of the leg and foot. Some of them are attached to a small ligament at the knee, called the patella—knee pan—which serves the purpose of a fulcrum over which the cord pulls in moving the lower limb and foot. These muscles are all advantageously disposed of for power, and are so arranged as to hold the limb

THE MUSCULAR SYSTEM. 53

erect, and to balance the body upon the upper portion of the limb.

136. *How much is it supposed that a muscle contracts, and what is the process of the contraction of a muscular fibril?*

It is supposed that muscles are capable of contracting about one third their length. This contraction or shortening of the fiber is owing to a change in the diameter of the component parts of an ultimate fibril. This action may be well illustrated by placing a dry rope in water: wetting it increases its diameter but shortens its length.

*Figure VII.*

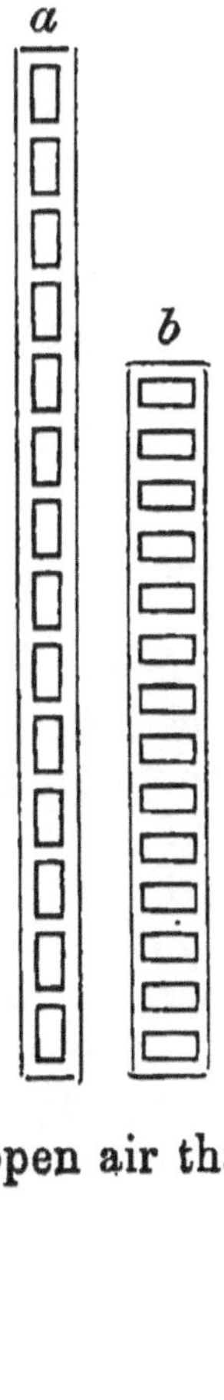

In *Fig. VII* the contraction of a muscle is illustrated. In *a*, the beaded portions of the fibrils are at rest. The diameters of those portions are greatest lengthwise of the fibrils. In *b*, the muscle is contracted, and the beaded portions have their longest diameters crosswise of the fibrils. The natural state of a muscle is what is called its *tonicity*, and it is supposed to be a constant strain or stretch. This keeps the muscle in a position in which it is always ready for action. Just what the element is that causes the motion of muscle is not yet decided. It is certain however that a good supply of oxygen and electricity in the blood greatly facilitates the action of the muscle, and so muscular exertion can be continued for a much longer time in the open air than otherwise.

137. Which is the largest tendon in the body?

It is the one which is attached to the calf of the leg, extending down to the heel, and which is used to raise the body upon the toes. It is called the tendon of Achilles, because the great Grecian warrior Achilles is said to have been killed by the wound of an arrow in this tendon.

138. What is found connected with the muscles?

Around the fibrilla of the muscle, but not entering into them, is woven a very beautiful net-work of very fine capillaries, communicating with arteries on the one hand and veins on the other, so that a plentiful supply of blood is constantly poured around the contractile elements of the muscle. Thus its exhausted energies are replenished, and its substance nourished. The veins receive the unappropriated blood, and conduct it back to the heart; and thus a continual stream of fresh arterial blood is poured through all the muscular tissue. By this means the vitality of the muscle is maintained. The involuntary muscles are even more abundantly supplied with vessels than those of animal life.

139. What other peculiarities are there in the arrangement and action of muscles?

Each muscle is provided with one or more antagonistic muscles, or those that produce motion in an opposite direction. The only exception is in a few muscles of the head and neck. When one set of these muscles are contracted, the others must relax, otherwise the body would not move. Medicines that produce nausea, or sickness at the stomach, will relax all of these muscles. Another peculiarity is, that all the component parts of a muscle do not contract at once, but one portion of the fibrils contract, then another, and another, and so on. If the whole muscle contracted at once, its action must necessarily be very short. While the muscle is contracted decomposition is going on, and the blood is shut out of that portion of the muscle; but as different parts of the fibrils take up the process of contraction, there is a partial building up of the muscular structure from the nutriment of the blood, even while contraction of the other parts is taking place.

140. What is the arrangement of the muscles of the alimentary canal?

As it is necessary in the performance of the general functions of the alimentary cavity, that there should be *motion*, as well as innervation and secretion, muscular fibers are everywhere attached to the back of the mucous membrane forming that cavity. The general arrangement of these fibers consists of two layers: the first composed of circular fibers, which surround the meat pipe, the stomach, and the small and large intestines, like rings, or sections of rings; the second layer is composed of fibers running lengthwise of the meat-pipe, stomach, and intestinal tube. By the contraction of the circular fibers, the size of the cavity is diminished. By the contraction of the longitudinal fibers the parts are shortened. By their combined action they give the parts an undulating motion. This muscular coat is stronger and thicker, in the meat-pipe and stomach, than in the small intestine, and stronger in the small than in the large intestine. In the rectum, or outward terminus of the intestines, the muscular coat becomes thicker and stronger. In the pharynx the muscular coat is composed of six constrictor muscles, the fibers of which form sheets which cross each other in various directions. By the action of these muscles, both the length and caliber of the pharynx are diminished. In the stomach the fibers are arranged in three different directions: longitudinally, circularly, and obliquely.

The muscular coat of the alimentary organs, and particularly of the stomach and small intestine, is more or less developed in power and activity, according to the character and condition of the food on which the person habitually subsists. That food calling for a proper amount of muscular action in the stomach and intestines increases their strength, while food of an opposite kind conduces to emaciation, and inactivity of those fibers, rendering the action of the stomach and bowels sluggish and feeble.

141. Do the muscles act at the greatest advantage, or are they so arranged that power is sacrificed to save time?

The attachment of muscles for the movement of the body is at a great disadvantage, as will be seen by looking at *Fig. VIII.* The lines, *a, b, c,* are designed to represent the arm; *a*, the shoulder; *b*, the elbow; *c*, the hand; *d*, the point of attachment of the muscles which are to move the fore arm and hand. If the muscle was attached from *a* to *c*, more weight could be lifted than with an attachment from *a* to *d*; but it would take more time to lift it as the muscle would have to contract more. And with the present contractile power of a muscle it would not

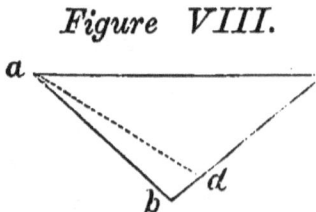

Figure VIII.

MUSCULAR ACTION.

contract sufficiently to raise the hand to the head. The muscles are usually attached near the joints, and as a hand near the hinge of a gate must use great exertion to move it; and as the hand near the hinge moves but little and slowly to move the farther end of the gate rapidly, so with the action of muscles, as illustrated above, they are arranged with reference to saving time, but it is an expenditure of force. All muscular action, as we have shown, is attended with a great waste of muscular fibril, so no more muscular exercise should be put forth than is necessary for health and to accomplish our necessary work. For this reason farmers, mechanics, and all laboring classes, should have their tools handy, and calculate their work to save all unnecessary steps and blows; let the "head save the heel." Houses should be arranged so that the woman's work will be handy, and all useless steps avoided. It will save the health of women who otherwise would be overworked. A person raising a hundred-pound weight in the hand, exerts a force on the muscle equal to eighteen hundred pounds. The act of leaning over and straightening up is a strain on the muscles of the legs of many thousand pounds. The muscles of some lower animals are much stronger in proportion to their size than those of man. A flea harnessed, will draw seventy or eighty times its own weight, while a horse draws but six times its own weight. The common beetle bug has been known to throw a weight placed upon it three hundred and twenty times heavier than itself.

142. What examples illustrate the rapidity of muscular movement?

Some persons have pronounced as many as 1500

letters in a minute, combined of course in words; but to do that, it would require the contraction and relaxation of one or more muscles to each letter, each contraction occupying not more than one-fiftieth of a second. The wings of some animals must move many thousand times in a minute to produce the humming which is heard while their wings are in motion. Some birds fly 60 feet in a second, while a race horse scarcely exceeds 40 feet in the same time. A falcon of King Henry II, flew from Fontainbleau to Malta in one day, a distance of about one thousand miles. *The precision* of muscular movement is seen in the rapidity with which a singer can accurately strike notes in any part of the scale. These sounds are all produced by contracting and relaxing the muscles of the larynx.

143. How may the muscles be strengthened, and kept in a healthy condition?

The muscles, to be kept healthy, should be used. Using the muscles increases the flow of the blood through them, and thus the waste, or decomposed particles, are carried off, and nutritive particles are placed in their stead. Plenty of exercise in the open air is imperatively necessary for a healthy condition of body in students, professional men, in-door mechanics, and females. Farmers who work in the open air, if not overworked, enjoy the benefits of the wholesome air of heaven. This exercise should be in a condition free from care, or burdening of the brain, thus giving the blood a chance to flow healthfully through all the system. These exercises in the open air are better, because then the oxygen is freely inhaled into the system, and thus through the blood the muscular system is built up. The air contains more oxygen in cold than in warm weather, and therefore greater muscular activity can be attained in winter than in summer, and for this reason, other conditions being equal, the body will gain in weight faster in winter than in summer.

144. What else is necessary in relation to muscular exercise?

Muscular exercise, or labor, is more conducive to health, and more of it can be endured, if it is done gradually, than if a violent exertion is made. After work we need repose. After the muscles have been used violently, or after vigorous exercise, the muscles should gradually be brought into a state of rest. Sleep is the grand restorative after severe muscular exertion; this alone gives back to the muscle its life and strength.

145. Mention some of the most healthful exercises?

Riding on horse-back, walking, climbing mountains, running up and down stairs, sawing wood, planing boards, rotary motions with both arms extended while the lungs are filled with air, or carefully moving the arms back till the backs of the hands touch if possible.

146. What should be the position of the body in standing or walking in order to properly develop the muscles?

The body should be upright, with the head, shoulders and hips thrown back, and the breast forward. Constant bending over will cause a round-shouldered, crooked, mean, diminutive appearance. But the appearance is the smallest evil. It causes the bones of the chest to press upon the internal organs of the body, and hinders their healthy action, causes short breathing, and pain in the chest, weakness of the lungs, and finally consumption. A person who stands erect, can stand with more ease, labor better, and travel farther in a day, than one who stoops. Students, when sitting at their studies, or in writing, should avoid a stooping posture. If we always keep the body in a proper position it will tend to make the back bone firm and strong. In all bodily or mechanical labor the body should be bent, or lean on the hip joints; the trunk should be kept as straight as possible.

147. What can we say, in conclusion, of the muscular system?

The muscles of the human system are a wonderful combination of flexible cords, by which, in an instant, the body, or any part of it, may be moved by the will in almost any conceivable direction.

148. What protecting coat invests these soft structures, and the delicate organs of the body?

They are everywhere invested by bandages, called *fascie*. They are composed of fibrous tissue of various thickness, and are divided into two classes, called *cellulo-fibrous*, and *aponeurotic*.

149. What of the *cellulo-fibrous* fascie?

It invests the whole body between the skin and deeper parts, and affords a medium of connection between them. It is composed of fibrous tissue, arranged in a cellular form. It affords a yielding and elastic structure, through which the minute vessels and nerves pass to the papillary layer of the skin, without obstruction, or injury from pressure.

150. What of the *aponeurotic* fascie?

It is strong and inelastic, composed of parallel

tendinous fibers, connected by other fibers passing in different directions. In the limbs it forms distinct sheaths, inclosing all the muscles and tendons constituting the deep fascia. In the palm of the hand and sole of the foot it is a powerful protection to the structures.

## Chapter Four.

### THE CIRCULATION OF THE BLOOD.

151. What are the organs used for the circulation of the blood?

The *Heart*, the *Arteries*, the *Capillaries*, and the *Veins*.

152. What is the central point of the circulation?

The heart is the grand central engine of the human body, which propels the blood to all parts of the system. It is situated in the chest, resting by its lower surface on the diaphragm, and somewhat to the left of the middle line of the body. It is suspended to the spinal column in the upper part of the chest by the blood vessels and ligaments connected with its upper portion. It extends downward, forward, and slightly to the left, behind the breast-bone. The place where it is felt beating being against its extreme left and lower point.

153. What is the heart?

The human heart is a strong, hollow, heart-shaped organ, composed of muscular fibers, disposed in several layers, so as to form fibrous rings and bands, which afford it the greatest possible amount of strength for its bulk. It has several compartments and valves. The fibers of which the heart is composed cross themselves in at least three directions, and many of these fibers join with each other in many places. The heart is lined with a continuation of the lining of the veins, and covered with a similar coat. It is encased in an enclosing membrane called pericardium. This contains a small quantity of a fluid like water, so that the heart actually floats in a liquid, and does not rest firmly upon any hard surface.

154. Is the heart a single or double organ?

The heart is a double organ; externally it appears to be a single organ, but internally it is double. It is externally about five inches in length, by three and a half in diameter, and weighs about ten ounces. Internally, it contains four compartments; two right, and two left. These compartments are called *Auricles* and *Ventricles*. These pairs of *Auricles* and *Ventricles* have not any communication with each other, therefore they are considered as two distinct hearts. The right is called the *venous* and the left the *arterial* side of the heart. The whole heart will hold nearly a pint.

155. What can you say of the *auricles* of the heart?

The right auricle is larger than the left; its interior has five openings, and two valves. The *auricles* are very similar. They are the uppermost cavities of the heart, and are smaller than the *ventricles*, or lower ones. The auricles have thinner walls than the ventricles, but their walls are thicker than those of the veins. They are capable of considerable dilation, so that the receptacle of the blood may be enlarged in case of a sudden effort of the body when the blood is sent in great quantities to the heart. Were it otherwise the veins would be in danger of rupture. These auricles are constructed of a continuation of the veins that terminate in them. They are the inner extremities of the veins. Their name is derived from an appendage or extension; that in the right auricle appears a little like a dog's ear.

156. What of the *ventricles* of the heart?

The ventricles are the lower cavities of the heart. Their walls being thicker than those of the auricles, they have more power of contraction, which is necessary, as they drive the blood from the heart. The right ventricle propels the blood only to the lungs, while the left ventricle sends it to all parts of the body except the lungs. The ventricles are similar, the right being a little more capacious, and its walls thinner, than the left. The inner surface of the ventricles is smooth, at the same time it is uneven, many bundles of fibers extending across the cavity, which gives them a spongy appearance.

157. What further arrangement is connected with the auricles and ventricles of the heart?

The auricles have several openings on their sides toward the veins, at which there are no valves, and one opening toward the ventricles, with valves, through which, in case of the right auricle, blood can flow back if much force is exerted; but in case of the left auricle none can flow backward, or the lungs might be endangered. There are only two openings to each ventricle, one into, and the other out of it; these are both situated at the upper part of the ventricle, and are both furnished with valves.

158. What are these valves?

The valves are muscular fibers arranged in such a form that the blood can pass through them; but the contracting of the ventricle presses the blood back against the valve and closes it so that the blood cannot pass back again. The same contraction of the muscles of the ventricle forces the valves open out into the arteries, and the pressure of the blood in the artery closes the valve again. Thus the ventricles by their alternate contraction and expansion, act very much on the principle of a force pump; sucking the blood into the heart from the veins and forcing it out again through the arteries.

159. What is the course of the blood into and out of the heart?

The venous or impure blood is received into the right auricle, and then passes into the right ventricle, which contracts and forces the blood into the lungs, where it is purified, and passes back into the left auricle. From the left auricle it passes into the left ventricle, which contracts and throws the blood through the arteries into all parts of the system except the lungs. This blood imparts nourishment to the bones, cartilages, ligaments, tendons, membranes, muscles, and nerves, and supplies the various secretory organs with the blood from which they separate their lubricating, solvent, and other fluids; and having served its purpose it returns again to the heart to be forced into the lungs, purified and again forced out into the system as above described.

160. What then appears to be the necessity of two sets of auricles and ventricles, or two hearts?

The necessity for two hearts arises from the fact that all the blood, as soon as it has made its circuit through the body, needs to be acted upon by the air, and must therefore be forced through the lungs; but the delicate structure of the lungs would not tolerate the force necessary to drive the blood through them and then out again into all parts of the body, or, if they would, such a force would drive the blood through them too rapidly; therefore, after the blood has been passed through the lungs, it is received by another heart, whose contraction throws the blood into all parts of the body.

161. What causes the alternate contraction and dilation of the heart?

It is the direct action of the nerves of organic life connected with the heart. When the blood passes into the right auricle of the heart, the nerves of the heart are immediately stimulated to action; force is communicated to the muscles, which contract, closing the upper valve of the auricle, forcing open the valve into the ventricle, where the presence of blood excites through the nerves the muscles of the ventricles, which contract, closing the valves into the auricles, and throwing the blood out of the heart. This is the constant action of each auricle and ventricle.

162. Is the action of the heart continual?

It is. The blood is the life of man, so it must be kept constantly flowing. If the heart should cease its work the blood would cease to circulate and we should immediately faint, or die. Both auricles of the heart contract at the same time, and this is closely followed by the contracting of the ventricles, which occasions the double beating which may be observed by listening carefully with the ear over the heart. It is similar to the beat and return in the ticking of a clock.

163. If the heart is continually in action when does it get any rest?

The contraction of the auricles occupies only one-fourth the whole time of a beat, so that the auricle rests the other three-fourths of the time. The contraction of the ventricle occupies one-half the time of the whole beat, so it rests the other half of the beat.

164. How many times does the heart contract or beat in a minute?

In health, at maturity, the heart beats from sixty to eighty times; more frequently in woman than in man. Early in life it beats oftener. In advanced years less frequently. The health, state of mind, and amount of exercise taken, also affect the pulse.

165. How much power is exercised by the human heart in its pulsations?

It is supposed that the left ventricle of the heart acts with a force equal to sixty pounds at every beat; or, at eighty beats per minute, the muscles of the human heart exert a force in one hour equal to two hundred and eighty-eight thousand pounds.

166. How much blood passes through the heart in an hour?

The capacity of each ventricle is between one and two ounces. At eighty beats per minute, if only one ounce is received and thrown out at each beat, it will equal five pounds per minute, or more than a barrel per hour.

*Figure IX.*

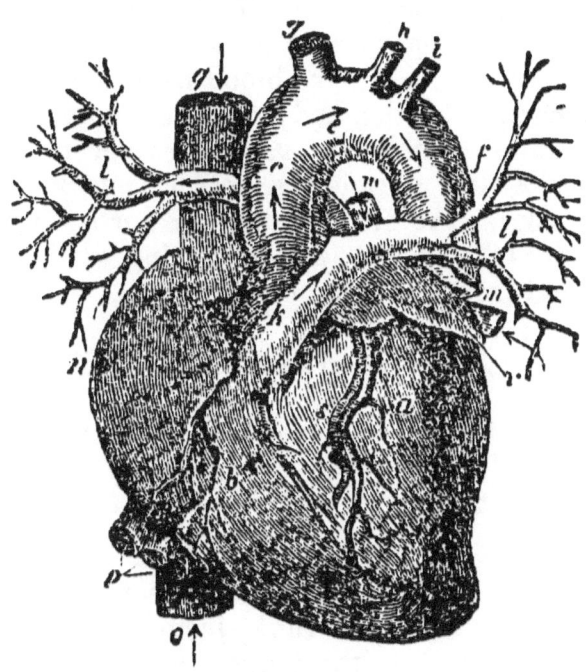

THE HEART.

167. What does *Fig. IX* represent?

*Fig. IX* is an external view of the heart. *a*, left ventricle; *b*, right ventricle; *c, e, f*, aorta, or the principal artery leading from the left ventricle. It is through this that the blood passes to all parts of the system.; *g, h, i*, are the internal and external arteries of the face, head, neck, arms, &c.; *k*, pulmonary artery; *l, l*, its right and left branches, which carry the impure or venous blood from the right ventricle into the capillary vessels of the lungs;

*m, m*, veins of the lungs which carry the blood back from the lungs to the left auricle of the heart; *n*, right auricle; *o, ascending vena cava*—the great vein which returns the impure blood from the lower portion of the body to the right auricle of the heart; *q, descending vena cava*, or the great vein which returns the impure blood from the head and upper portion of the body to the right auricle. These unite to form the *vena cava*, or *returning hollow ; r*, left auricle; *s*, left coronary artery, which carries the blood from the aorta into the substance of the heart; *p, portal veins*, which return the blood from the liver.

### ARTERIES.

168. How is the blood carried from the heart to the different parts of the body?

Through the arteries.

169. What are the arteries?

They are the set of tubes that commence from each heart as a single trunk, from which they extend, in one case to the lungs, in the other they divide and subdivide till they reach all parts of the system. They are dense, tough, cylindrical tubes, which form they retain when emptied of blood, and even after death. From this circumstance the ancients regarded the arteries as air vessels.

170. How many coats or layers of membranes compose the arteries?

Three: the outer coat is of sinewy fibers; the middle is muscular, being made up of contractile fibers; the inner is like the lining of the heart and veins. It is nervous; or, in other words it is a membrane through whose substance are interspersed the nerves of organic life. The outer coat is firm and strong; the middle is thick and soft; and the internal thin and polished. The arteries maintain a cylindrical form unless forcibly compressed.

171. What is the particular benefit of the muscular or elastic structure of the arteries?

It allows the arteries to distend when the heart forces the blood into them, and, when its action intermits, the arteries return to their former size, which propels the blood forward, and thus the action of the arteries themselves is like that of compressed air in fire engines; so that arteries not only lead the blood to the different parts of the system, but they force it along.

172. What is the number of the arteries, and what their capacity?

They number about one-third as many as the veins in the system generally, and their capacity is about one-half that of the blood vessels in the lungs; but as the arteries receive the blood directly from the heart, it is forced rapidly through them.

173. Where are the arteries placed?

The arteries are buried deep in the flesh, and this serves two purposes: first, its warm current is less exposed to loss of heat; and secondly, they are less likely to be tapped by superficial injuries.

174. How can it be known in case of a wound that an artery has been severed?

When an artery is cut, and not closed by the effect of the injury, the blood flows in jets, corresponding with the pulsations of the heart. It will flow in such a case with serious rapidity, and life will be lost, unless the flow is speedily stopped by forcible compression. In case an artery is severed, if possible, a bandage, handkerchief, or cord, should be drawn tightly over the artery, between the wound and the heart. This bandage should be twisted tightly by the aid of a stick inserted beneath it while loosely tied. A knot may be made in the bandage and placed over the artery, or a smooth stone, a chip, or a few pennies may be placed under the bandage to produce pressure directly on the artery. An elastic suspender, or something of that kind, makes the best bandage, as it does not wholly check the flow of the blood. Blood is essential to the vitality of the parts immediately around the wounded part, and it would be better to lose a little blood than to have none received below the wounded part. There are cases when the flow of blood can only be stopped by taking up and tying the artery. If you find your efforts to check the blood failing, immediately secure a good surgeon to assist you.

175. How do the arteries of the system arise?

They arise from the left ventricle of the heart by a single trunk called aorta, or air-keeper, that turns down with a beautiful arch, from which branches sweep out into the arms and lead directly up to the head. The main subdivisions above and below are few and large, as may be seen in *Fig. X*. Throughout their entire length, however, they give off numerous small branches and twigs.

176. How many classes of arteries are there in the greater or systemic circulation?

There are twenty-three classes, and these are all used to convey the pure or arterial blood through the body.

*Figure X.*

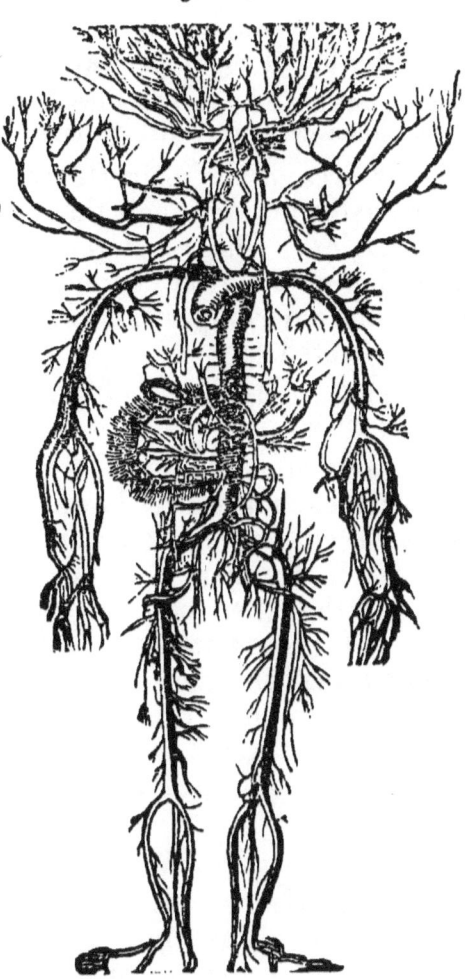

THE ARTERIAL SYSTEM.

177. What is noticeable in *Fig. X* concerning the principal branches of arteries?

That those which start immediately from the *aorta*, or main artery, are at right angles with it, which moderates the impetus of the blood; but those branches toward the extremities of the arteries leave the main branch at an acute angle, which will aid the blood to a more rapid flow there.

178. Is there more surface covered by the branches of the arteries combined, than in the main artery, and why?

The combined area of the ultimate terminus of the arteries is vastly greater than that of the main trunk of the arteries. This arrangement allows a more quiet motion of the vital current in the extreme vessels, where decomposition and reforming of structure is effected.

179. Is there any connection between the arterial tubes in the various parts of the body?

Yes; in all parts of the body the arterial tubes communicate with each other by branches passing between them, called inosculations, and these connections increase in frequency as the vessels diminish in size, so that their final distribution is a complete circle of inosculations. This arrangement provides against obstructions, which are liable to occur in the smaller branches of the arterial tubes. When one of these branches is obliterated another branch above enlarges and makes up the loss in the circulation.

180. What are the pulmonary arteries?

They are the arteries which carry the impure blood into the lungs. The right pulmonary has three branches, which carry the blood to the three lobes of the right lung. The left pulmonary is the largest division; it passes to the root of the left lung. The pulmonary arteries divide and subdivide in the substance of the lungs, and finally terminate in a net-work of capillary vessels around the air cells and passages of the lungs.

## CIRCULATION OF THE BLOOD.

181. How is the substance of the lungs nourished?

The pulmonary arteries which convey the blood from the heart to the lungs, and the veins which carry the blood back from the lungs to the heart, act no part in nourishing the substance of the lungs. But the bronchial arteries, and their corresponding veins, extend to every portion of the pulmonary structure; with these is connected a capillary system of muscles, and nerves of organic life, which preside over the process of building up and keeping in repair the lungs themselves.

182. Do the arteries empty directly into the veins?

They do not; but they empty into what is called the capillary system, which is an extremely minute net-work of vessels and nerves from which the veins arise.

### CAPILLARY SYSTEM.

183. What are the capillaries?*

They are minute vessels, intermediate between the arteries and veins. They are not a *distinct* system, but merely fine tubes by which the arteries are continued into the veins.

184. What is the size and number of the capillaries?

They are of a uniform size, very regular in the distribution of their branches, and their tubes are from one five-hundredth to one six-thousandth of an inch in diameter. It is supposed there are more than one thousand of them to every square inch of the human body. There is no place where you can even prick the body with a pin but you will

---

*The word capillary signifies a hair, used here to represent hair-sized vessels.

pierce some of these capillaries, and blood will flow from them.

185. What function of the body is accomplished in the capillaries?

The capillaries are the most essential parts in the circulation. In the capillary vessels all the organic functions take place. Through them the blood is brought into very near relations with the parts upon which it is to act. It is in the capillaries that the perfect nutrition of the structures takes place: worn-out material is separated, and new material is consolidated. In its passage through the capillary vessels the blood loses its vivifying properties and the florid color it received in the lungs, and becomes dark, impure, and charged with effete matter, resulting from the decomposing particles of tissues. All the waste matter not collected into the excretory passages of the several organs are carried along by the capillaries into the veins, to be purified in passing through the liver and lungs.

186. How is the circulation of the blood effected in the capillaries?

The circulation in the capillaries seems to be to a great extent independent of the heart's action, and to be regulated by the organic nerves which preside over the functional process, and distribute the blood to the various parts as needed. The sum of the diameters of the capillary vessels greatly exceeds that of the arteries and veins, which enables the blood in them to move slowly, and even sometimes take a retrograde motion, which greatly helps in the building-up process of the body.

187. In what parts of the body are capillaries found?

In *all* parts of the body, except the outer coat

of the eye, the tendons, and the nails. They not only serve for the important process of secretion and nutrition, but in them a portion of the animal heat is produced.

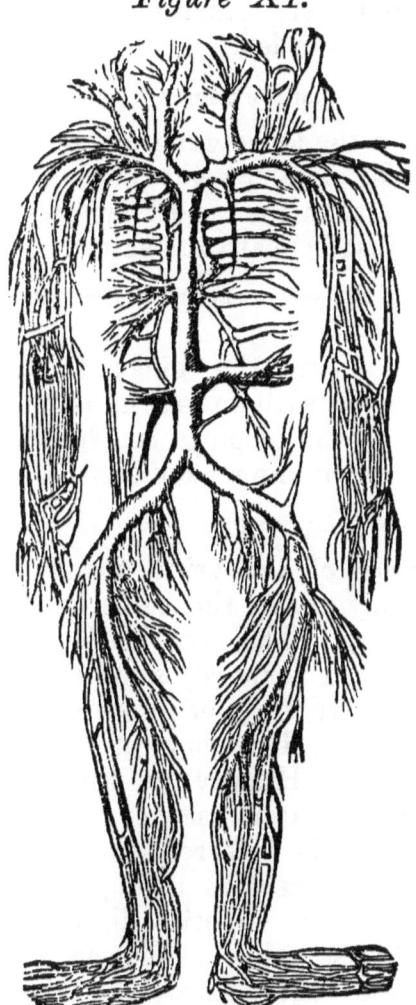

VENOUS SYSTEM.

*Figure XI.*

188. What is *Fig. XI* designed to illustrate?

The principal veins of the human system.

189. What are the veins?

The veins are the vessels which return the blood to the heart, after it has been circulated through the various structures of the arteries and capillaries.

190. What is the structure of the veins?

They have three coats similar to those of the arteries; but they are much thinner than the arteries, and collapse and flatten on becoming empty.

191. Where do the veins commence?

The veins originate in small radicles, or veins, in all the textures of the body. These empty into larger trunks, those into larger still; the sum of all the radicles being much larger than the main trunk, which facilitates the flow of the blood to the heart, and as the blood is continually coming into a larger channel from small extremities it flows faster and faster, which is just the reverse of the action of the arteries.

192. What is the general course of the veins?

The larger veins follow the same general course as the large arteries. The smaller veins, however, are much more numerous than the small arteries, and are most abundant just beneath the skin.

193. What other difference can you mention between the veins and arteries?

While the arteries are elastic, the veins are not. The veins are also furnished with valves in them, which facilitate the flow of the blood to the heart, but almost entirely prevent its flow backward.

194. What are the valves of the veins?

In the lining membranes of the veins, pouches or bags are formed. They act in the same manner as valves in machinery. The free border of the valvular flap is directed forward, allowing a free current toward the heart, but not in an opposite direction. It was the discovery of these valves in the veins which led Harvey, an English physician, in 1628, to the greater discovery of the circulation of the blood. He inferred that the blood could pass in but one direction through the veins, and consequently in an opposite direction through the arteries. These valves are most numerous in the veins of the extremities. The use of these valves is to prevent the blood, when pressure is made upon the veins, from flowing back into the capillaries. The position of some of these valves can be seen on the back of the hand. By rubbing the skin down with the finger over the vein, the blood is pressed against the valves, distending the veins. If the finger is carried below the valve, the blood is pressed away from it, and the vein between the finger and valve will be empty.

195. How many classes of veins are there?

There are three classes: *sinuses, deep,* and *superficial.* The sinuses are excavations in the structure of an organ, and lined by the internal coat of the veins. There are several of these sinuses in the *dura mater* of the brain. These run in various directions on the inside of the skull, and most of them empty into the great veins of the neck. They afford a free passage of blood from the brain, even if by excess of arterial action this organ should be overcharged with blood.

196. What of the other two classes of veins?

The superficial are found near the surface, just under the skin, while the deep veins usually accompany the arteries in pairs, one on each side, and these two courses of veins frequently communicate with each other. They at last unite at the auricle of the heart, into which they pour the blood drained from every part of the body.

197. What force carries the blood through the veins?

The agencies which are employed to carry the blood through the blood-vessels back to the heart are not so well understood as the action of the arteries. It appears, however, aside from that help derived from the suction of the heart and the aid of the vein-valves, that the blood is helped on its way through the veins by the action of the muscles about the veins, and temporary pressure upon the surface of the body; so light rubbing occasionally of the surface of the body must facilitate the circulation of the blood. The affinity of venous blood for oxygen is supposed by some to cause it to urge its way on to the heart and lungs. It is probable that the great principle moving it onward is the direct controlling influence of the organic nerves, which are everywhere present in these veins.

198. What are the pulmonary veins?

They are the veins which return the pure or arterial blood from the lungs to the left auricle of the heart. These veins are four in number, and they differ from the veins in general, in being but little larger than their corresponding arteries, and in accompanying, singly, each branch of the pulmonary artery.

199. What are the veins which are called the *Portal System?*

It comprises those vessels which receive their blood from the intestinal canal, the stomach, the spleen, &c. These small vessels unite into a larger trunk, and, instead of passing directly to the heart, they form what is called the portal vein, which passes into every part of the liver, where the blood is again collected by a series of vessels which convey the blood to the heart. By this means the blood which passes from the bowels, &c., to the heart, is strained, and prepared to enter into the general circulation; with the other venous blood, it is forced into the lungs, and from the heart out again through the system.

200. What are the component parts of the blood?

If the blood be examined microscopically when

first drawn, it appears to be made up of a transparent liquid called serum, or plasma, and a number of circular bodies, mostly of a red color, called *corpuscles*, or minute bodies. The material of the blood is albumen, fibrin, and several salts, some of which are found in distinct crystals. When the blood is exposed alternately to the action of oxygen and carbonic acid, these red corpuscles lose their circular form, and become star-shaped, and are finally destroyed; it is calculated that millions of them are thus destroyed at each pulsation of the heart.

201. What is the color of the blood?

The blood if drawn from an artery is of a bright red color; but if drawn from a vein it is purple. This color is caused by the impurities in the blood. The odor of the blood is the same as the breath of the animal from which it is taken.

202. Are there any other substances connected with the blood?

Yes; the waste substances produced by the action of organs, and the nutritive material adapted to replace the waste.

203. How much blood is there in an ordinary-sized human body?

From three to four gallons. Of this, from one-fourth to one-third is supposed to be in the arteries; from two-thirds to three-fourths in the veins; a large proportion of the whole being in the arterial and venous capillary vessels. The blood weighs some 30 pounds in a person weighing 150 pounds.

204. How long does it take for all the blood to make a circuit through the system?

This depends on the health of the person, and the amount of exercise taken. It is supposed that an amount of blood equal in weight to all in the body makes its circuit in from three to eight minutes. The blood is thrown in an instant from the heart to the extremities, but its passage back, as already shown, is slower.

205. How much blood passes through the heart in twenty-four hours?

Not far from ten thousand pounds, or five tons, equal in weight to about seventy men.

206. What is necessary to a healthy conditon of all parts of the body?

A healthy circulation to all those parts. If the system is supplied with pure blood, and that circulates properly to every organ, it is health.

207. What is absolutely essential to promote a proper circulation of the blood?

It is necessary to pay special regard to the clothing, to see that the limbs and arms, hands and feet, are properly and sufficiently clothed. If the clothing is too scanty on these parts the blood is chilled, which greatly retards its circulation; the efforts of the heart are also increased to carry the vital current to the suffering parts, which is a needless and unnatural wear of that organ. It is important also that the clothing should be worn loosely upon the body, as much pressure upon the body prevents the rapid flow of the blood from the extremities to the heart. Many instances of persons suffering with cold feet might be almost if not entirely cured by wearing loose boots or shoes of sufficient thickness, instead of such thin and delicate tight ones as some now wear.

208. What can you say of proper food, as affecting the circulation, and the building up of the system?

As the blood is made *from* and *of* the alimentary substance contained in our food, and the quality of that aliment depends on the quality of the food eaten, so, if we would have our blood properly freighted with suitable nutriment, and our body kept in a healthy condition, it becomes a matter of the highest importance that we give the most scrupulous care and attention to the proper selection of our food. And also to select that kind of food which our system can most readily digest and assimilate to its own uses. Otherwise we cause a useless wear of the system, and thus shorten our lives.

209. Are there many diseases of the circulating organs?

These organs, though constantly in use, and easily excited by muscular movement and mental emotion, are affected with but few diseases. Many diseases, called disease of the heart, are merely sympathetic affections, and the real disease and difficulty lies in other organs. Heart diseases are most common late in life, at about sixty years of age. Those persons who have any tendency to disease of the heart, either real or sympathetic, should be on their guard against sudden exertions, and, to as great an extent as possible, avoid mental anxiety, grief and alarm, as these all tend to increase such difficulties.

## Chapter Five.

### THE LYMPHATICS.

210. *What are the lymphatics?*

They are minute, transparent vessels, uniform in size, having various valves. They constitute what is called the absorbent system. They are named from the substance they convey, lymph, a watery fluid, which they gather and pour into the blood.

211. *What is the office of the lymphatics?*

Their office is to collect the nutritive products of digestion from the alimentary canal, and the effete, disorganized matter from all parts of the body, and convey them into the venous blood. They also have the power of absorbing substances applied to the skin. Green leaves of tobacco applied to the abdomen will often produce distressing sickness. Poisons thus taken up by the blood from the surface of the body, being undiluted by the juices of the gastric cavities, pass directly into the circulation, and are therefore more powerful than when modified by passing through the internal absorbent vessels. In some cases where the passage from the mouth to the stomach has been closed by disease, nutriment has been infused into the system by means of a bath of warm milk. Shipwrecked sailors in an open boat have slaked their thirst by wetting their clothing in salt water, or what is better still, by being wet in a rain storm.

212. *What is the origin of the lymphatics?*

They originate in a delicate net work, and are

distributed throughout the skin and the various surfaces and internal structure of organs. There is scarcely a part in the whole body where these lymphatic vessels are not found, but, in some parts they are so extremely small that they cannot be discovered without the aid of a microscope. In the brain, where they are supposed to exist, they have not as yet been discovered even by the microscope. They are remarkable for their uniformity in size. They are of a knotted appearance, and very frequently divided into two nearly equal branches. They proceed in nearly straight lines toward the root of the neck.

213. What is connected with these lymphatic vessels?

They are intercepted in their course by numerous oblong bodies, called lymphatic glands. These are small, oval bodies, of a reddish ash color, and vary in size from one twentieth of an inch to an inch in diameter. They are somewhat flattened, and are larger in young persons than in the adult, and are smallest in old age. These glands are situated in different parts of the body, but abound mostly in the thorax and abdomen. Leaving these, the lymphatics converge from all parts of the body so as to pour their contents into tubes, which open into large veins leading to the heart, near the bottom of the neck.

214. What is the construction of the lymphatics?

Like arteries and veins, they are composed of three coats, frequently connected together, and having valves. It is these valves that give to the lymphatic vessels their knotted appearance. These valves are most numerous near the lymphatic glands. The lymphatics are smallest in the neck, larger in the upper extremities, and still larger in the lower extremities.

215. What are the lacteals?

They are the lymphatic vessels of the small intestines, which convey the milk-like fluid called chyle to the thoracic duct. These are the nutritive absorbents, and they are connected with the numerous glands of the mesentery.* It is in the small intestines that most of the alimentary absorption of the body is effected.

216. What is the difference in the functions of the lacteals and the lymphatics proper?

The lacteals convey nutritive or new matter into the mass of blood, to replenish the tissues; the lymphatics take up and return to the blood the surplus nutrient materials, and also old or waste particles, in order that they may be used in the secretions of the body or got rid of at the excretory outlets. The function of the lacteals is called the absorption of composition, that of the lymphatics the absorption of decomposition. The lymphatics proper, pervade, as before shown, the whole body, arising in great numbers from the external skin, from all the internal membranes, vessels, and cavities, and from the substance of all the organs. But the lacteals arise only from the mucous membrane of the alimentary canal, and principally from the mucous membrane of the small intestine. There is, however, no difference in the structure of a lymphatic and that of a

---

*The *mesentery* is a broad fold of what is called the peritoneum. It is six inches in length and four inches in breadth. It is fastened to the back, and serves to retain the small intestines in their position. It contains between its layers the mesenteric vessels, nerves, and glands.

lacteal; but one elaborates chyle, and the other lymph.

217. How many kinds of lymphatics are there?

Two: the superficial and deep. The superficial follow the course of the superficial veins, and they join the deep lymphatics. The glands of these superficial vessels are placed in the most protected positions, as in the hollow of the ham and groin, and on the inner side of the arm. The deep lymphatics accompany the deep veins.

218. What is the principal center of the lymphatic system?

The thoracic duct, or *recepticulum chyli*. It commences in the abdomen, and ascends through the diaphragm to the root of the neck, and then turns forward and downward, pouring its contents into the venous blood just before it enters the right auricle of the heart. The thoracic duct is some eighteen inches in length, and in size about equal in diameter to a goose-quill. Its termination is provided with valves to prevent the admission of venous blood. Before emptying its contents into the blood, it receives several large branches, or trunks, from the lacteals, and the lymphatic branches, from nearly all parts of the body. It is the common trunk of all the lymphatic vessels of the body, except those of the right side of the head, neck, and chest, and right upper extremity, the right lung, right side of the heart and the outer surface of the liver. These empty into the right lymphatic duct, and this empties into the venous blood-vessels of the right side of the chest near the heart.

219. Are the lymphatics the only absorbent vessels?

No; the radicles, or small veins, perform a

very important function in the stomach, by the rapid absorption of the watery portion of all liquids placed there. It thus conveys them to the general circulation without their passing through the circuitous route taken by the food.

220. What can you say in general terms of the lymphatic system?

It is an appendage to the venous system, furnishing it with all the assimilating materials by which the body is nourished, as well as conveying to it the effete susbtances which are to be eliminated from the vital domain. These two systems are connected at several points, and the structure of the lymphatic vessels much resembles that of the veins. The venous capillaries and the lymphatics appear, to some extent, to reciprocate in function, and the lymphatics always empty their contents into the veins. In the lymphatics, as in the arterial and venous systems, the nerves of organic life are distributed, and they preside over all their functions. In the lymphatic vessels some of the most important changes take place.

221. What is chyle?

It is a liquid substance, composed of digested food, and is prepared for nutrition in the mesenteric glands. It is of a milky-white color, and is of the same chemical composition, whatever may be the food from which it is formed. It is not, however, of the same vital quality. That formed from animal food, when taken from the body undergoes putrefaction in three or four days, while that selected from vegetable food will resist decomposition for several weeks.

222. What is lymph?

It is a watery fluid, differing from chyle in its color, being almost colorless, and differing also in the fact that the lymph is made mostly of decomposed matter, while the chyle is always formed from new matter. This lymph thus formed, is mostly, if not entirely, waste matter, and thrown off through the excretory ducts.

223. Are the lymphatics and lacteals the only organs that absorb nutrition?

No; in the mucous membrane of the lungs and stomach, the thin fluids are taken up by the veins.

224. What is the general view of the absorption in the system?

The extremities of veins act as absorbent vessels, taking up the greater portion of useless, injurious, or worn-out matters; the lymphatic vessels return the unused or surplus nutritive matter; they also serve as helpers to the veins when they are obstructed, or their task imperfectly performed. The elements of the blood in the capillary system pass through the coats of these vessels and undergo chemical, vital changes. Such elements as are needed, repair the waste and build up the structures of the body. Other elements are separated and carried back into the circulation, to be changed or thrown off.

## Chapter Six.

### THE NERVOUS SYSTEM.

*Figure XII.*

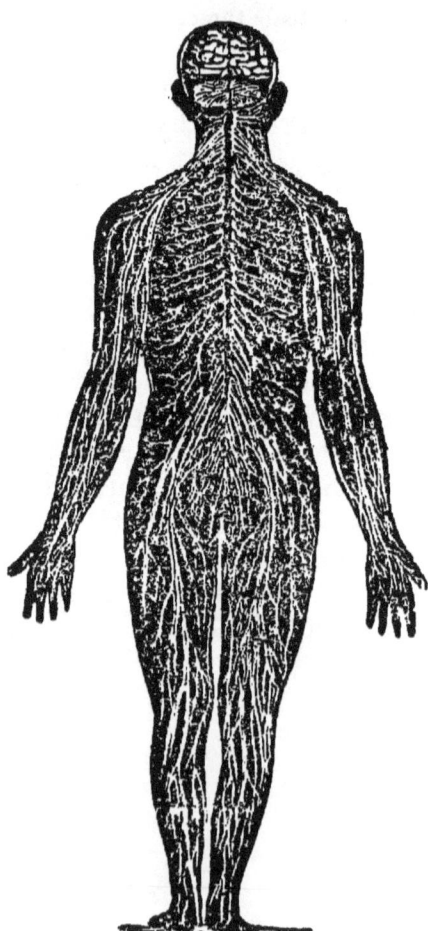

NERVOUS SYSTEM.

225. What is the nervous system?

It is, in many respects, the most important and interesting portion of the human body. It is the highest order of organized matter, is the immediate organism of vitality, and the vital operations, and the intellectual manifestations; and hence it has been said, that the nervous system constitutes *the man;* and that the bones and muscles, and the whole assemblage of internal organs, with their various functions, are only intended to sustain and serve the nervous system. All impressions on the mind from the external world, and all mandates from the will to the muscles, are conveyed through the medium of the nervous substance. All motions, changes, or functional actions which are performed by the muscles of the body, depend on the power, energy, or influence transmitted to the muscular tissue from the nerves.

226. Is vitality peculiar to the nervous system alone?

The nerves are more highly endowed with vital properties and powers, than any other substance of the body; and they are, in the animal kingdom, the immediate instruments of vitality in all the wonderful operations of the body. Vitality in various degrees, however, pervades all the tissues of the living body. The blood seems to be a living fluid, as also the chyle, especially in its more advanced stages of assimilation. Through the action of the nerves on the appropriate organs the food is digested into chyme, and thence into chyle, and thence into blood. The blood is transformed into the various solids and fluids of the system, and, at the same time, by the nerves, it is supposed that the temperature of the body is regulated. By virtue of the vital endowments of the nerves, we perceive our internal wants, and external conditions, and relations, and by these nerves we act upon the muscles, and through them upon the bones in our voluntary motions. By the mysterious endowments of the nervous substances, we think, and reason, and feel, and act, as intellectual and moral beings.

227. Do the functions of all organized bodies depend on a system of nerves?

There are, in all organized bodies, both animal and vegetable, a class of functions which are concerned in the nourishment, growth, temperature, and general sustenance of the body as an organized being. There is a tissue in vegetable bodies, which, in its functional character, corresponds with the nervous tissue of animals, as nearly as the functions of vegetables and animals correspond in their processes and results. The vegetable seed, by virtue of its own vitality, excited to action by a genial soil and other appropriate circumstances, puts forth its little roots into the earth, and absorbs foreign matter, and converts it into the substances and texture of its own organism. So far as those vital operations are considered by which chyme, and chyle, and blood are produced, and the blood circulated through the system, and the body in all its parts nourished, and its growth and development effected, and all the other functions of organic life sustained, the animal differs but little from a vegetable; and in health, is equally destitute of animal consciousness. In animals, however, there must be care used in the proper selection of substances to nourish the body, instead, like the plant, of its getting all its nutrition on a fixed spot. So, in animals, there are organs of sensation, locomotion, and prehension, subject to voluntary control. The primary office of these organs is to perceive and procure the materials by which the body is nourished, and place them within reach of those organs of nutrition. The Zoophytes, the lowest order of animals, have furnished a matter of controversy with naturalists, as to whether they were animal or vegetable. They are but dimly conscious of their being, and are nourished by means which hardly demand faculties superior to those with which the vegetable is nourished.

228. Into how many systems may the nervous system be divided?

Into two: the *organic*, comprising those nerves concerned in, and presiding over, the functions of digestion, absorption, respiration, circulation, secretion, organization, or the process of structure, and the production of animal heat. The other system is called the *cerebro-spinal*, comprising the nerves of sensation and motion. To this latter belong consciousness, the perception of external impressions and internal affections, reflection, volition, and other faculties called intellectual. The first of these systems is composed of all those nerves called *sympathetic*, which preside over the functions of organic life. The second comprises the brain, spinal marrow, and nerves of sensation and motion. This might also be divided into other systems called the *motory*, or nerves of voluntary motion; the *sentient*, or nerves of sensation; and the *mental*, or the brain.

229. What is the structure of the nerves?

The nervous substance is sometimes white, sometimes gray, and in the organic system of a reddish color. This tissue of the nerve fiber is enclosed in membranes or sheaths. The ultimate nerve fiber is tubular, consisting of an external, thin and delicate membrane, which forms a sheath, within which is contained a more opaque substance, called the *white substance of Schwam;* and within this white substance is a transparent material which may be made to move in the cavity of the tube.

230. What is the comparative size of the component parts of a nerve?

The nerve fibers vary in size from one two-thousandth to one fourteen-thousandth of an inch in diameter. A nerve is made up of a bundle of these fibers, enclosed in a sheath.

231. What is the grand center of the nervous energy of the body?

The nervous system has been by many reckoned as the brain and spinal marrow, with numerous cords, branches and twigs, dispersed over the whole organized system. Those holding this theory claim the brain as the grand center of the nervous system, or a kind of *galvanic battery* which continually generates nervous energy and presides over all the vital functions of the system. The brain is undoubtedly the central point of sensorial power, but it does not seem to be the presiding center of those nerves by which the development of the different parts of the body is effected. If the brain were the presiding center of vital operations in the formation of the body, then all the branches belonging to this center would issue from it, and go out with the blood-vessels, to preside over their functions, in the formation of other parts, and enter into the texture of parts thus constructed; but, instead of this, the branches of the spinal nerve are mostly distributed to the voluntary muscles, and to the outer surface of the body. Again, children have been born with all parts of the body well developed, except the brain and spinal marrow, and it is a fact that the brain and spinal marrow are among the last parts of the body brought to that consistency which enables them to exercise a functional power. Although, as we shall show, the condition of the ceerbro-spinal system of nerves may and does greatly affect the healthful action of the organic nerves, yet we conclude that the grand center of nervous vital energy is the central point of the nerves of organic life.

232. What is the central point of the nerves of organic life?

In the midst of those parts of the body first produced, in its natural order of development, we find a mass of nervous matter, which, in composition, very nearly resembles the brain. This is undoubtedly the first-formed portion of the human body, and is the grand center which presides over all the functions concerned in the growth of the body and the functions of nutrition during life. This central mass of nerve is situated at the roots of the diaphragm, in the upper and back parts of the abdominal cavity, or nearly back of the pit of the stomach, and consists of several parts; two semi-circular bodies about an inch long and half an inch broad, lying, one on the right, and the other on the left side of the back bone. These are called the SEMILUNAR GANGLIA. They are probably at first one, and afterward partially separate to accomodate themselves to the duplicate arrangements of the body. They are, however, closely connected by many large branches, which pass from one to the other, and form what is called the SOLAR PLEXUS.

233. What do we find connected with the *solar plexus?*

Numerous branches of nerves go out from this central brain, in different directions. They are threads of a reddish color, and have connected with them oval bodies called *ganglia*, which are never so large as peas. These ganglia and nerves extend along each side of the spinal column from the atlas to the coccyx, and distribute branches to all the internal organs and viscera, and communicate with all the nerves of the body. The branches which are given off to the internal organs accompany the arteries to the same.

234. What other arrangement is connected with these nerves?

The ganglia above mentioned serve as smaller and subordinate brains, which become the more special centers of development, and of perception and action, to individual organs, or particular apparatuses of organs; and these special centers, in their turn give off numerous branches, some of which enter into the texture of the blood-vessels formed for, and appropriated to, their service in the construction of their particular organs; others are distributed to the contractile tissue or muscles of those organs to convey the stimulus of involuntary motion; others are the conductors of impressions made upon the organs, to their special centers. To establish an intimate connection between the different special centers, and bring them all into direct relation to each other, and to the common center, large cords run directly from one center to another; and numerous branches go from each center, to interlace and unite and form plexuses,* with branches coming from several other special centers, and from the great common center. Some of these plexuses are formed around the internal organs and are named after the arteries extending to those organs; as the mesenteric, hepatic, splenic, &c. To the naked eye the plexus looks as though the nerve was lost, but by the microscope each fiber can be traced distinctly through the plexus.

---

*Plexus is a Greek word signifying to weave. The plexus is a net-work of nerves.

## The Nervous System.

235. What can you say of the order and development of these organic nerves, and their presiding centers?

As the alimentary canal and the other organs associated with it in the general function of nutrition are developed before any other part of the body, their special centers and nerves are the first produced. At an early stage in the development of the body numerous fibers rise on each side of the general mass, which form a pair of large cords, called the trisplanchnic nerves, which give rise to an uninterrupted series of small brains, which gradually separate in a longitudinal direction, and draw farther and farther apart, keeping up their connection with each other by intermediate branches, till they form a connected range of about fifteen little brains, on each side of the spinal column, from the diaphragm to the top of the neck. The trisplanchnic nerves are divided in their upper portions into from three to seven or more branches, which terminate in as many of the little brains in the two ranges. Eight or nine more of these little brains are arranged in a similar manner from the diaphragm downward. So, in the completely-developed body, there is a continued series of these brains, or special centers, on each side of the backbone, from the base of the cranium to the inferior extremity of the spinal column. Each of these little brains sends out numerous branches, which serve to unite the little centers to each other; others plunge into the muscles; and others form connections with the nerves and muscles of animal life. Other branches go to interlace and form numerous plexuses with branches of others from the same, and of the opposite side, and from those more deeply seated among the viscera, and from the great central mass itself. From these plexuses, again, numerous branches are given off to the different organs, entering intimately into their texture. And all the branches and twigs of this system of nerves, as they proceed along their course to their destination, cross, and unite, and divide, and interlace, so as to form of the whole system one extended net, the meshes of which become smaller and smaller, as the nerves approach their inconceivably-minute termination in the organs. These two series of little brains, with their connecting cords, &c., bring all the organs with which they are connected into a very close union, and establish between them a most powerful bond of sympathy. By the aid of the numerous branches which pass from the several plexuses to different organs, the whole assemblage of organs concerned in the functions of organic life, is, as it were, woven into one grand web of nervous tissue.

286. *What further can you say of this system of nerves?*

They preside over all the vital functions in the development and sustenance of the body, while the other special centers are more immediately concerned in the development of the organs employed in the general function of nutrition. It is probably the truth in the matter that the two series of brains or special centers which extend the whole length of the spinal column, are more immediately concerned in the development of the spinal nerves, and of the cerebro-spinal system generally, and of the other parts pertaining to the trunk and extremities.

237. *How many orders of ganglia of organic life are there?*

There are two : called the central, and the peripheral or limiting ganglia. The central are those connected with the internal organs, and are supposed to preside, generally and specially, over the functions concerned in nourishing and sustaining the body. The limiting are those which form the two ranges on the sides of the spinal column, and have been supposed to be more particularly appropriated to the general sympathies of the internal system, and are accordingly called the sympathetic nerves. This general system of nerves is called the ganglionic system. They are most commonly called THE NERVES OF ORGANIC LIFE, in contra-distinction to the brain and spinal marrow, with their branches, &c., which are called THE NERVES OF ANIMAL LIFE.

238. What is the effect produced by suspending the action of the organic nerves?

A single moment's entire suspension of the functions of the nerves of organic life, would be a death from which there would be no resuscitation. The functions of the nerves of the cerebro-spinal system may be suspended for a considerable time, and still the common vitality of the body be preserved. *Andrew Wallace*, a revolutionary veteran who lived to the age of 105 years, was struck down by lightning while tending a cannon on the fourth of July, soon after the close of the American Revolution, and lay seventeen days in a state of suspended consciousness. He revived after it, and was remarkably vigorous and active.

239. How many orders are there to the nerves of organic life?

Three: the first, those nerves that enter into the texture of blood-vessels, and go with them to their most minute terminations in the different tissues, and preside over all their functions of absorption, circulation, secretion, structure, &c.; the second, those nerves that go to the contractile tissue, or muscles of involuntary motion in the texture of the organs, and convey to them the stimulus of motion; the third, those which convey to the special centers, and, if necessary, to the common centers, the impressions made upon organs. The cords which serve to connect the special centers to the common center, and to each other, are probably composed of filaments of all these three orders.

240. What is observable in the distribution of the nerves of organic life?

The heart seems to require and possess but few of these nerves. This is likewise true of the large blood-vessels. But in the capillary system, or minute extremities of the vessels, where all the important changes take place, the nerves much more largely abound. Of all the organs of the body, the stomach is the most remarkable for its nervous endowment, and sympathetic relations. Lying near the great ganglionic center, it receives a large supply of nerves directly from that source, and is thereby brought into the closest sympathetic union with the common center of organic life, and through it, with all the organs and parts in its domain. By the arrangement and distribution of plexuses, also, the stomach is brought into very direct relations with the heart, liver, lungs, and all the other organs.

241. From whence do the organic nerves derive their nourishment and support?

They evidently derive their support, as well as

the elements by which they operate to control and regulate the organic functions, in a great measure directly from the arteries, for which purpose also, some of their structural parts penetrate the arterial coats?

242. *Is the mind conscious of the operations of the organic nerves?*

In a healthy condition of the bodily organs these nerves have no sensibility of which the brain takes cognizance. For instance, the brain does not feel food in the stomach, nor blood in the heart, nor air in the lungs, nor bile in the liver, yet their pressure is recognized or felt by the organic nerves. When, therefore, we are *conscious* that we have a stomach or a liver, from any feeling in those organs, we may be certain that something is wrong. These nerves are endowed with an exquisite organic sensibility, which qualifies them most perfectly for the performance of their constitutional functions in the living system; and the complete integrity of those functions essentially depends on the healthy properties of the nerves. But the organic sensibility of these nerves may, by continued or repeated irritation, become exceedingly diseased, and a diseased sympathy may be induced and permanently established. In this state of things, all the functions of organic life are necessarily impaired. The food is less perfectly digested in the stomach, the chyle is less perfectly elaborated, the blood necessarily becomes deteriorated, and the whole system, in every part and tissue, consequently suffers. By continued irritation, inflammation may be induced, and the most painful sensibility developed in these organs. This state of things is not only distressing, but is always injurious, and often hazardous to life.

100 PHYSIOLOGY AND HYGIENE.

*Fig. XIII.*

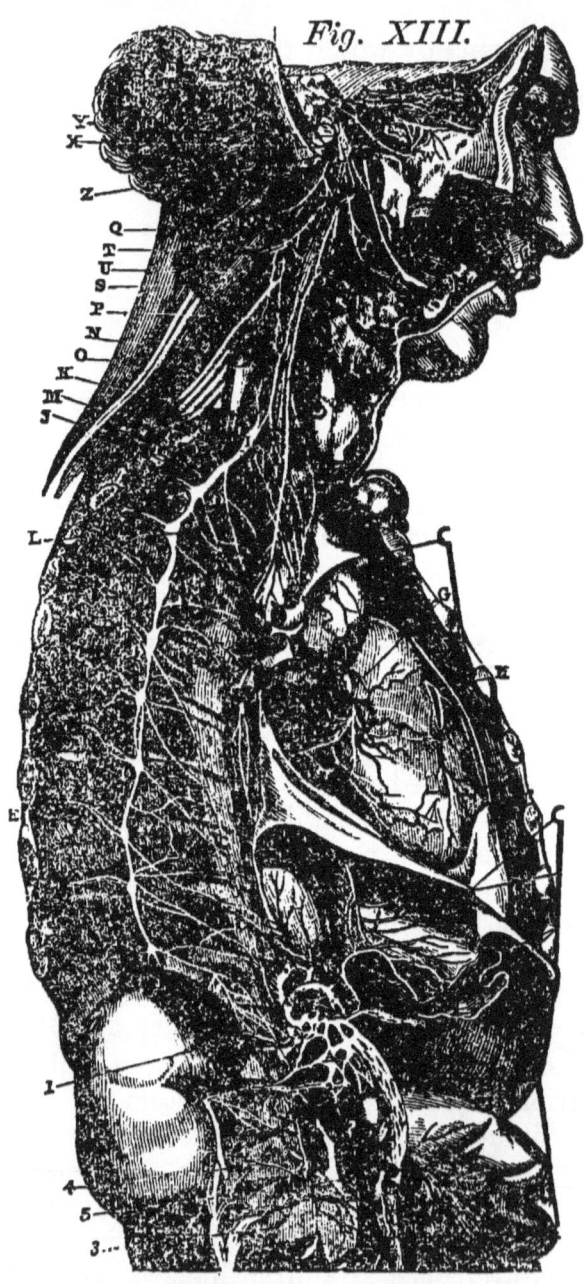

GANGLIONIC SYSTEM.

243. What is *Fig. XIII* designed to illustrate?

The *ganglionic* or *organic nervous system*. As may be seen by the little lines running in every direction, and to every organ of the body, it forms a complete system of telegraphic communication with every part of the body. A A A, is the *semilunar ganglion* and *solar plexus*, the great presiding center or brain of the organic nervous system. Various sensations usually referred to the heart, have, no doubt, their source in this ganglion. B, and C, C, greater and lesser *trisplanchnic nerves*. These nerves connect with the nerves of the base of the cranium, with the ganglia of the vertebral column before described, with the *solar plexus*, and with the *renal plexus*, which extends from the solar plexus to the kidneys. D D D, are the *thoracic ganglia*, which consist of twelve ganglia on each side, resting upon the head of the ribs. E, their internal branches which follow the pulmonary artery, to the plexus of the throat, heart, and splanchnic nerve. Several nervous cords from the lower ganglia, as will be seen in the figure, unite to form the splanchnic nerve. F, internal branches of the thoracic ganglia, which extend to the roots of the spinal nerves. G, right *coronary plexus*, which is placed around the right coronary artery of the heart. H, left *coronary plexus*, around the left coronary artery of the heart. These are both connected by branches with the principal nerves of the ganglionic system. I, J, K, L, M, N, O, P, Q, R, S, T, superior, middle, and inferior, *cervical glands*, with their branches. They extend from the base of the skull to the third cervical vertebra, and send branches to the organs of the throat, upper part of the chest, especially to the heart.

U, *submaxillary ganglia*, which are connected with the glands of the lower jaw and several of the facial nerves. W, *naso-palatine* branch of the vidian nerve, extending to the nose and palate. X, *spheno-palatine* branches of the vidian nerve. These are four or five nerves, extending to the mucous membrane, spongy bones of the nose, and to the pharynx. Z, *auditory* nerve, extending to the ear. 1, *renal plexus*, extending from the solar plexus to the kidneys. 2, 3, 4, *lumbar ganglia* with its branches, which consist of four ganglia on the front part of the lumbar vertebra, which are distributed to all the viscera of the lower organs of the body. 5, the *aortic plexus*, which is a combination of the lumbar ganglia around the aortic artery in the abdomen.

244. In a close view of the nerves of organic life, as illustrated in the above figure, which organ seems to be placed in a position to have the most sympathetic influence on the others?

The degree of sympathetic influence which each organ has on the others, is always proportionate to the functional importance of the organ in the system, and the nearness of its nervous relation to the great center of organic life. The stomach holds an immensely-important station in the assemblage of vital organs. It is supplied largely with nerves from the great center of organic life, and associated by plexuses with all the surrounding organs, and hence it sympathizes more directly and powerfully with every other internal organ, and with every part of the living body, than any other organ; and, in turn, every other internal organ, and every part of the living body, sympathizes more directly and powerfully with the stomach than with any other organ. If proper and healthful food be placed in the stomach, it is healthfully excited, and all the other organs rejoice with it and take hold with alacrity to perform their labor; but if an improper substance irritates the stomach, all the other organs mourn with it, and their functions are disturbed by it. By carefully considering this bond of sympathy in the entire domain of organic life, we shall certainly realize the force of St. Paul's expression, when, using the human body as an illustration, he says: "And whether one member suffer, all the members suffer with it, or one member be honored, all the members rejoice with it."*

245. If there is no sensibility of which the mind takes cognizance in the action of the organs of the vital domain, how do we know when we are diseased, or when we want food or drink?

It was the *healthy* action of which we said the mind had no knowledge. We have just stated that unhealthy and irritating action produces the most distressing agony. This the mind is cognizant of, not because of the nerves of organic life, but because of another class of nerves which are established for the express purpose of imparting to the mind a knowledge of the wants and ills of the human system. This system of nerves, with the intimate connection between the two systems, and their effect upon one another for good or ill, will now occupy our attention. This second system of nerves, as we have already told you, is called

THE CEREBRO-SPINAL NERVOUS SYSTEM.

246. What is the grand center of the cerebro-spinal nervous system?

---

*1 Cor. xii, 26.

## The brain.

**247. Where is the brain situated?**

The brain is in the head, occupying the whole inner part or cavity of the skull, but separated from it by a thin membrane.

**248. Of what is the brain composed?**

Of the same substance as the nerves. It resembles marrow, and is filled with blood-vessels, the whole brain having a grayish color.

**249. How large is the brain in a grown person?**

It is about six inches long, five inches wide, and four inches thick. It weighs from three to four pounds, and will fill the two hands of a man. The brain of man is larger than that of any of the lower animals except the elephant and whale. The elephant's brain weighs about eight pounds, the whale's brain about five pounds.

**250. Has a person more than one brain?**

Yes; there are two brains. The larger one is called the cerebrum, and another, about one-half as large, below and behind it, is called the cerebellum. These two brains are equally divided into two parts by a deep cut or separation, reaching nearly through them from front to back. These halves of each brain are precisely alike in shape, and together form a pair of brains, just as we have a pair of eyes and ears. We see the wisdom of the Creator in thus arranging the brain organs in pairs. One side of the brain may be injured, and the other will perform the functions of those organs. Or one side may be paralyzed, and yet the life of the person is not destroyed, but still the brain and nerves act.

## Figure XIV.

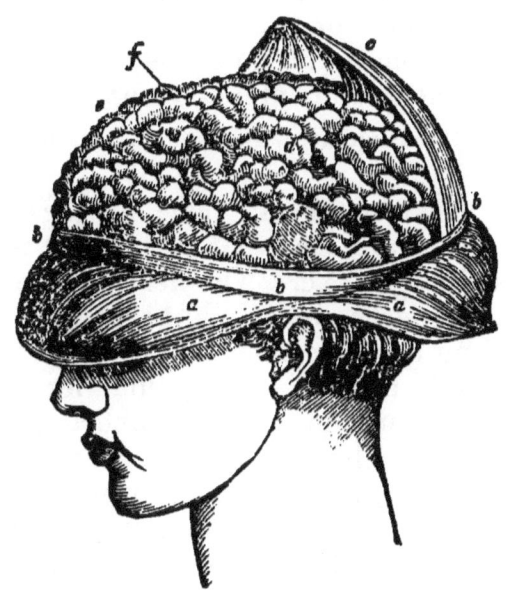

**251. What does *Fig. XIV* represent?**

It represents the brain exposed, showing the external surface of the cerebrum, or large brain. *a, a,* is the scalp turned down; *b, b, b,* the cut edges of the skull bones; *c* represents the *dura mater*, or membrane enclosing the brain, lifted up by a hook, exposing the brain; *d* represents the left hemisphere of the brain; *f* represents the deep cut or fissure between the right and left sides of the brain, which divides the organs of the brain into pairs, as above stated.

**252. Does all the substance called the brain lie in the head?**

No; the animal nerves are only small strings of the same substance, running from the top of the head to the very extremities of the body.

253. Which is the largest nerve in the body?

The spinal marrow, which is situated in the center of the spine or back bone; extending from the middle of the brain, down between the arms, through the neck and spinal column.

254. Does this large nerve send out smaller ones through the body?

Yes, in great numbers; some of these nerves also extend from these great nerves to the ears, eyes, nose, tongue, &c. The nerves give us feeling; without nerves we should be without feeling. If we had no nerves connecting the eye with the brain we could not see. We could not smell or taste if there were no nerves connecting the nose and tongue with the brain. If a nerve is cut or destroyed the organ to which it is attached loses its function entirely. If the nerves connecting the hands with the head were destroyed, our hands might be burned in the fire and we would have no consciousness of pain or suffering.

255. How many coverings are there to the brain?

There are three. The first or outer covering is the *dura mater* of the brain; see *a, Fig. XIV.* It is a strong, whitish, fibrous membrane, which adheres to the internal surface of the cranium, or skull. From the internal surface of the dura mater, portions extend inward to support and protect different parts of the brain, and externally, other processes extend outward for sheaths for the nerves passing out of the skull and spinal column. The second membrane is the middle covering; it is very thin and transparent. It surrounds the nerves until their exit from the brain. The third is the *pia mater*, and is the internal covering, consisting of numerous blood-vessels held together by thin layers of tissue. It invests the whole of the brain and each of the windings, by extending through all the fissures between them. The pia mater is the nutritive membrane of the brain. It is through its blood-vessels that arterial blood is distributed, which is received from the arteries entering the front and back parts of the skull.

256. What part of the brain is called the cerebrum?

That portion which is seen in *Fig. XIV*. It is divided into the right and left hemispheres by the line *f*. Each of these hemispheres is divided on its under surface into front, middle, and back lobes. The surface of the cerebrum, as you may see, presents a number of slightly-convex elevations.

257. What is the cerebellum?

The cerebellum contains about one-sixth or one-seventh part of the brain. It is pear-shaped, and is attached to the cerebrum by a band of thick fibers. It lies directly beneath the cerebrum in the back of the head.

258. What seems to be the order of development of the cerebro-spinal nerves?

1. The spinal nerves, commonly described as those arising from the spinal marrow, but which *probably* preside over the formation of that nerve. 2. The spinal marrow itself. 3. Those ganglia of the brain which are essential to the functions of taste, smell, hearing, and sight, together with the special nerves by which these functions are performed. 4. The ganglia which constitute the mental and moral faculties. 5. The cerebral hemispheres themselves.

259. What is the spinal marrow?

The spinal marrow is that soft substance which lies in the hollow of the back bone, and is composed of the white and gray substances. It is naturally divided longitudinally, into a right and left half; each of which consists of a front and back column, so that the whole marrow is composed of four columns, or rather of two corresponding pairs; as the two front portions correspond with each other in form and character, and the two back ones correspond also with each other. They constitute a double spinal marrow, and give to each half of the body an independent existence so far as the spinal marrow and its nerves are concerned. And for this reason it is, that one whole side of the body may be paralyzed, while the other remains in full possession of its powers. The spinal marrow is enveloped in three different membranes, corresponding to the three enclosing membranes of the brain. Connected with the spinal marrow, through small intervertebral openings formed for the purpose, on each side of the spinal canal, are thirty pairs of nerves, which are called the spinal nerves. A portion of the filaments which compose each spinal nerve rise in the back portion, which are the nerves of animal sensation, some of these go to the muscles of voluntary motion and convey to the mind information concerning the motion of those muscles, and thus enable the mind to regulate their motion. The rest run to the outer skin of the body, and are the nerves of sensibility or feeling on its surface. Branches of these nerves run to the fingers' ends, where they are highly sensitive to the touch. The filaments of nerves that arise from the front portion of the

## THE NERVOUS SYSTEM. 109

spinal marrow are nerves of motion. These convey the stimulus of motion, in obedience to the *will*, to the voluntary muscles, causing them to contract. Although these filaments start from opposite sides of the spinal marrow, they unite into one cord, almost immediately on leaving it, and go in one cord to their muscles, yet these filaments may be traced distinctly by the anatomist on dissecting the cord. The spinal marrow seems to be a connecting link between the brain and the various nerves of the body, or, as it were, a protected thoroughfare through which the mandates of the *will* may be carried to the body, and information carried back to the mind.

*Figure XV.*

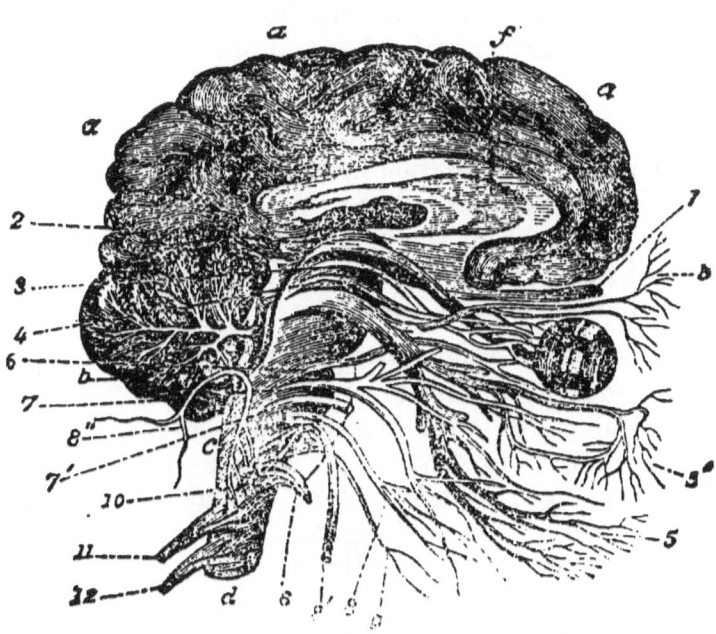

CRANIAL NERVES.

260. What is the *medulla oblongata?*

The spinal marrow, *d, Fig. XV*, passes upward through an opening in the base of the skull, extending about an inch into the cranium. The head of the spinal marrow is divided into six parts, or three pairs of bodies. Two on the front called the pyramidal bodies. Two behind, called the restiform bodies. Two at the sides called the olivary bodies. Besides these, a strip of medullary matter, which is the origin of the several nerves particularly associated in the function of respiration. These pairs of bodies are united in a single bulb, about one inch in length, and about two-thirds of an inch in diameter, and is commonly called the medulla oblongata. From the sides of this bulb, rise, as may be seen in the above figure, several pairs of nerves, and from its top arise all the other parts within the cranium.

261. What is the *arbor vitæ?*

When either lobe of the cerebellum is cut, it presents a tree-shaped arrangement of marrow-like matter, called arbor vitæ. It is seen in *Fig. XV*, just back of the medulla oblongata. A gray mass in the trunk of this tree, with saw-like edges, is called the *corpus dentatum*.

262. What is the arrangement of the ganglia of the brain?

By looking at the brain from below, we begin with the medulla oblongata (see *Fig. XV*), to the back part of which are added the cerebelli, from the right to the left of which, around the upper part of the oblongata, extends the *pons*, above which other ganglia are placed. Some of the fibers from the cord nerves extend into them all, and into the cerebelli, while the remainder extend upward and forward into the great ganglia.

The cerebri are, so to speak, folded back, over, upon, and by the side of, the other parts, slightly overhanging the cerebelli, but quite covering in the central ganglia. Where the surfaces of the cerebri come in contact with the central ganglia, they do not adhere to them, but both surfaces are free, and moistened with the same fluid, as if they had not been thus placed. These places where they thus come in contact are called ventricles. There are several of these ventricles in the brain. A dissection of the brains of inebriates frequently discovers these ventricles filled with alcohol. In dropsy of the brain there is a collection of watery fluid in these ventricles. Sometimes, however, the water is outside the brain, in which case the skull may be tapped by a skillful surgeon. The most dangerous cases are when the water is in the ventricles.

263. How many pairs of nerves are embraced in what are called the cranial nerves?

There are nine pairs of them, all of which, in *Fig. XV*, are marked numerically, as they are named. 1 is the *olfactory* nerve, the nerve of smelling. Nerves pass off from this which are distributed upon the mucous membrane of the nose. 2, *optic* nerve, the nerve of seeing. It comprises two large cords, extending from the medulla oblongata. At its front end it extends into the nervous membrane called the *retina*. This nerve is always present where the faculty of vision exists. 3, 4, are nerves of motion connected with the eye. These go to the muscles which serve to roll the eye, and direct the pupil toward the object of vision. 4, is used to give a pathetic expression to the eye, and hence is called the *pathetic* nerve.

5, with its branches, is called the *trifacial* nerve, and is distributed to every part of the face, forehead, eyelids, lips, nose, jaws, and ears. It unites freely with the facial, and several other nerves of the head, and with a great many twigs of the nerves of organic life. It communicates with the organs of all the five senses, and brings all the parts with which it is connected into a direct and powerful relation with the stomach and the whole domain of organic life. This nerve is connected with all the nerves of the teeth; this explains why decaying teeth may give rise to head-ache, ear-ache, &c. This nerve being connected with the pneumogastric nerve of the stomach, both are affected in that distressing malady sick head-ache. This, you will see, is the largest nerve of the cranial group. 6. These are also nerves of motion connected with the eye. These nerves when paralyzed, cause squinting. 7. The seventh pair are *facial* and *auditory* nerves, and are connected with the chin, lips, angles of the mouth, cheeks, nostrils, eyelids, eyebrows, forehead, ears, neck, &c. It is on these nerves that the expressions of the face depend. 8. The eighth pair of nerves consists, as may be seen, of three classes of nerves; *glosso-pharyngeal*, *pneumogastric*, and *spinal accessory*. The glosso-pharyngeal, or *tongue-and-pharynx* nerve, is distributed to the mucous membrane at the base of the tongue, to the tonsils and mucous glands of the mouth, and to the throat. When this nerve is paralyzed the voice is destroyed, and the act of swallowing hindered. The pneumogastric, or *lungs-and-stomach* nerve, forms connections and plexuses with almost every nerve in the region of the throat, neck, and thoracic cavity, to such an extent that it has been called the middle sympathetic nerve. It sends branches to the pharynx, or top of the meat-pipe; to the larynx, or organs of voice at the top of the wind-pipe; to the wind-pipe in all its branches and whole extent. It sends branches to the plexus of the heart, to the plexus of the lungs, some twigs to the solar plexus, and to the plexuses of the liver and spleen. But the main body of this nerve descends to the stomach, and is distributed over that organ, uniting extensively with the nerves which come from the solar plexus. The spinal accessory seems to be connected with the muscles essential to the production of voice, and is also connected with many branches of the spinal nerves. The ninth pair of nerves regulate and control the muscles of the larynx, or organs of voice. It is a nerve of motion to the tongue, controlling in a measure its use in speech. It is connected with the eighth pair, and many other nerves of the upper portion of the body.

264. What are the spinal nerves, and how many pairs are there?

The spinal nerves are so called because they take their origin from the spinal cord. There are thirty-one pairs of them; they furnish the principal nerves to the trunk, back, and extremities. The spinal nerves are divided into *cervical, dorsal, lumbar,* and *sacral*. The cervical consists of eight pairs, connecting with the upper extremities. The dorsal has twelve pairs, which are connected with the back, chest and abdomen. The lumbar nerves are five pairs, and the sacral nerves six pairs. These are connected with the lower portions of the body, the thighs and legs. There are also what are called sacral nerves, which are connected with the leg, and foot. These are in their divisions nerves of sensation and motion to the legs and feet. These spinal nerves convey impressions made at the surface of the body, including the extremities, to the brain, and transmit impulses to the muscles from the brain.

265. What is the great function of the brain?

It is the organ or instrument of the mind, and that by which the mind performs all its operations. It is the seat of all the intellectual and reasoning faculties of man, such as memory, hope, love, hatred, ambition. The brain is the seat of all sensation and knowledge, and the mind obtains its knowledge of all outward objects by impressions concerning them being conveyed to the brain through the medium of the nerves of sight, hearing, smelling, tasting, and feeling.

266. What is the effect of injuring the brain?

An injury of the brain will immediately affect the whole system. If the skull is fractured and depressed upon the brain, the person is immediately unconscious, having no thought or sense of objects around him. Instances are on record, where the brain has been injured, the person being insensible for several days, and when they were restored again to consciousness, they would speak the remainder of the very sentence they were speaking when they received their injury. If the brain is very seriously injured death will at once ensue.

267. Is the brain active during a state of profound sleep?

It is not. A person may be touched while in a state of sound sleep and not perceive it, or be conscious of it. When persons dream, they are not in a profound sleep. Dreams are many times in two parts. First, an earnest effort is made to get to some place or to gain some object; then comes a state in which you have no knowledge. Afterward your dream goes on again, with this difference, however: you find yourself at the place, or in possession of the object you were seeking, but hardly know how the feat was accomplished. This break in your dream is the unconscious state of your sleep, or that time when you have no dreams. That sleep is generally the most refreshing to the body, in which we dream the least.

268. What persons can get along with the least amount of sleep?

Those whose diet is of fruits, grains, and vegetables, and whose habits of life place them most of the day where they exercise in the open air. Persons of studious habits, or who follow in-door labor, require more sleep than those above mentioned. Those who eat but two meals a day, their other habits being right, enjoy their sleep the best, and can get along with a less amount of sleep than those who eat meat, or who eat their three meals a day. John Wesley, with an active nervous temperament, and a vegetable diet, performed extraordinary labors, with only from four to five hours sleep out of the twenty-four, while Daniel Webster, with a more powerful, but less active organization, and with the ordinary mixed diet, slept eight or nine hours.

269. What is the proper amount of sleep and the best time to secure it?

The great majority of those who have attained to long life were those who slept at least *eight* hours. The best period of sleep is to retire not long after dark, and to be up with the first rays of morning light. In the cold season, when nights are long, more sleep is required. All persons should make it their rule to retire if possible as early as half-past eight in the evening, and sleep as long as the slumber is quiet, if it be nine hours. Dreamy, restless dozing in the morning is generally more debilitating than refreshing. Those persons who indulge in the use of animal food, or eat gluttonously of any food, or who use spirituous liquors, or tobacco, are in danger of oversleeping, even to producing stupidity of mind, and indolence of body. Sleeping after a meal is always pernicious.

270. What should be the condition of the room, beds, and bedding, to secure refreshing sleep?

Sleeping apartments should be large, high, and well ventilated. The windows and doors should be so arranged as to allow a free circulatian of air, even night air. If the sleeping room is dark or damp, it should be occasionally dried with a fire in the room, but the fire, except in the case of very feeble persons, should be entirely extinguished and the room well aired before retiring in it for sleep. If practicable the sun should be allowed to shine in sleeping rooms some portions of the day. Ventilation of sleeping apartments should not be carried to an extreme. The air should not be allowed to blow directly on the sleeper, but there should be an opening somewhere by which fresh air from out-of-doors can be admitted into the sleeping apartment. In the most-severely-cold weather, say a crack in the window of a couple of inches, and in the warmest weather a door of the room open, and a third or half of the window open. The beds should be of straw, corn husks, or hair.

In case of those who are tender, they can use over this bed, a light, thin, cotton mattrass. No bed should be soft enough for the body to sink into it. Cotton or hair is much better for pillows than feathers. The bed clothing should be as light as possible, consistent with comfort. Linen or cotton sheets are better than flannel. For outside bedding, thin quilts are best in summer, and flannel blankets in addition for winter. The position of the body in sleep should be perfectly flat and horizontal, with the head a little raised; one common-sized hair pillow is generally sufficient. Healthy persons of correct dietetic habits may sleep at pleasure on the back, or gently reclining to one side. All however should carefully avoid reclining nearly on the face, or crossing their arms over the chest, as that brings the shoulders forward, contracts the chest, and materially affects the breathing. Placing the arms over the head in sleep is a pernicious practice.

271. Does a proper use of the mind strengthen the brain?

It does. The more the brain is exercised, if not overtaxed, the more firm and vigorous will be the operations of the mind; but if the brain is permitted to remain inactive, it will lose its healthy state, and all the operations of the mind must in consequence be dull and sluggish.

272. What protects the mass of brain from jars?

In early life the elasticity of the frame renders other protection against jars of the brain unnecesary, but as life advances, in addition to the increasing quantity of marrow in the bones, the *arachnoid* membrane beneath the brain increases in strength by an addition to it of sinewy fibers, which grow between the arachnoid and pia mater of the brain. These are filled with a fluid, and the brain rests upon it as easily as a person lies on a water bed. This cushion has become in old age, in some instances, an inch thick.

273. How is the surface of the brain arranged?

It is arranged in various winding elevations, which constitute the phrenological organs of the prevailing system of mental philosophy of the present day. As the brain is divided into right and left hemispheres, so all the organs of the brain are double. All phrenologists regard the cerebral portion of the brain as the seat of all the mental and moral powers, and the cerebellum as the seat of the sexual impulse. The cerebellum is also regarded as the generator of nervous influence to the muscles of locomotion. The whole brain, though the seat of sensibility, is itself wholly insensible. Any part of it may be cut, pricked, torn, or removed, without producing pain.

274. What is the mind of man?

It is the aggregate of all the functions of the brain. These are mental powers. The mental powers may be distinguished as faculties and propensities. The faculties together constitute the intellect. They are the powers concerned in thought and the formation of ideas, the thinking and knowing part of the mind. The faculties are divided into perceptive and reflective. The perceptive take cognizance of individual things and their mechanical properties, and are the functions of observation. The reflectives arrange, compare, and analyze subjects, and trace out their relations of cause and effect. These are the reasoning organs. The propensities are the feeling organs. They are the impulses, emotions, or passions, which impel us to action. The intellectual faculties devise means, seek out objects, and study methods to gratify these feelings or propensities. When the faculties have discerned the object, or ascertained the manner of satisfying the impulse or propensity, the will determines its instrumentalities—the bodily structures—to act in relation to its possession or enjoyment. Mind then, consists of faculties and feelings, or affections and thoughts.

275. Are the phrenological organs alone, a true index of moral character?

*No.* The cerebral portion of the brain, called the organs of the propensities, hold a more immediate relation to the physical condition of the nerves of organic life than do the intellectual faculties; so whatever increases the direct influence of the domain of organic life on the cerebral organs, proportionately increases the influence of the propensities over the intellectual and moral faculties. As the organs of destructiveness, combativeness, acquisitiveness, and amativeness, hold more near and special functional relations to the organic system of nerves, they are more readily excited by the excitement and irritation of the nerves of organic life. This being the case, it is more difficult for the intellectual faculties to weigh correctly evidence presented to it, and to arrive at conclusions of truth, and for the moral faculties to preserve their functional integrity in an excited condition of the organic nervous system, than when they perform only their healthful functions. So to form a correct knowledge of character from the organs of the brain, the state of the organic nervous system must be inquired into, as well as the general habits of the individual, as these all tend to modify greatly the character of the person. Cases are on record of persons whose phrenological developments alone would indicate them as among the best members of society, but whose intemperate habits so excited their destructive animal organs that they were mere beasts, seeking to destroy their best friends. All excess in stimulation will thus injure the mind, in proportion to the extent of organic nervous irritation.

276. What is true happiness?

All true happiness is that condition of mind which is the result of right feeling. The healthful exercise of all the mental powers is the condition to secure right feeling. Health of body and health of mind is happiness. A healthy condition of body is essential to health and strength of mind, while a healthy condition of the mind is happiness. This condition is one in which all the organs and propensities are in subjection to the man, and he governed in his course by right principles. The person then recognizes the hand of his Creator, and is led by the healthy action of his moral faculties to render to God due homage. Many, failing to see the connection between mind and body, attribute all their mental depression to the power of the Devil, or their own sins; yet these conscientious souls cannot tell really what these great crimes are. If they viewed matters in their true light, they might save themselves from despair. While realizing the goodness of God, and his tender mercies over all his works, and while trying to do every duty made known to us, and realizing our own feeble condition of body, we should learn to attribute a larger measure of our disconsolate feelings to the depressing power of disease, and less of it to the special frown of God upon us.

277. How and why are mind and body each affected by the condition of the other?

Because of that close sympathetic connection between the brain and all other parts of the system, or between the nerves of animal and organic life. Although, as before said, in health, the animal nerves have no direct control over the functions of those nerves that preside over the building up of the system, yet, there is such a sympathetic connection between them, that any violation of the healthy action of either affects the other. Excitement of the mind, or violent passion, affects the whole domain of organic life, and in some instances death is instantly induced. Such excitements and irritations frequently repeated lead to change of structure in the organs, and hence to disease. While the nerves of organic life are preserved in a healthy state, the mind is serene and cheerful, as in healthy childhood; but when these nerves are deranged, we are unhappy, we know not why. We long for relief, we know not from what. We would go, but we know not where. We would cease to be what we are, yet we know not what we would be. We look around for the cause of our grief, but in vain; we cannot find it, and conclude it must be God's displeasure for our crimes. This feeling is indulged until despondency like the pall of death enshrouds us, and envelopes us in its myriad folds. The brain and spinal marrow, and in fact all the nerves of the body, are nourished by blood-vessels over whose functions the nerves of organic life preside, so it is evident there is a close connection between the two systems.

278. What seems to be the connecting link between these two systems of nerves?

As before stated, the center of the system of animal nerves is the brain, and as we see by *Fig. XV*, the top of the medulla oblongata is the grand point from which these nerves all proceed, and the seat of the sensorial power in the system. The solar plexus is the center of the system of organic nerves. The pneumogastric or lung-and-stomach nerve, which passes directly from one of these centers to the other, and forms plexuses, and connections with so many of the organs of the vital domain, seems to occupy a middle ground between the nerves of organic and animal life. This is the nerve which establishes a most powerful sympathy between the brain and stomach. This nerve is probably the greatest connecting link between the two systems.

279. What organ of the body is most readily affected by the mind and by derangement in the organic nervous system?

The *stomach*, from its connection with the organic life center, and with almost all parts of the body by the pneumogastric nerve, sympathizes more directly and powerfully with every other organ than any other part of the body. So, for the same reason, every other part sympathizes powerfully with the stomach. Chronic indigestion impairs the functional power of the external skin. Excessive heat or cold on the surface, on the other hand, impair digestion. The most powerful sympathy exists between the brain and stomach. Intense and protracted, or excited and impassioned exercise of the mind, affects all the functions of the organic domain. It causes a sensation to be felt in the epigastric center. This sensation is usually referred to the heart, but the stomach more than any other organ is the seat of it. It is in a great measure through the stomach that other organs are affected by mental influences. Derangement of the stomach affects the liver, intestinal tube, and other internal organs.

280. *What most readily affects the condition and powers of the mind?*

The condition of the stomach and alimentary organs. The worst cases of insanity result from a deranged state of the organic nervous system, especially the stomach and intestines. This deranged state of the nerves of organic life constantly calls up in the mind improper thoughts and conceptions. The brain all the while may be in a perfectly healthy condition. That the real seat of insanity is in the nervous system instead of the brain may be seen in the fact that many instances are on record of insanity when the brain itself was not diseased except sympathetically. So also instances are cited where large portions of the brain were diseased and no derangement ensued. Again, cases where persons with debilitated stomachs were thrown into a state of derangement by eating a meal of pickled cucumbers. One of the greatest causes of insanity is loss of sleep.

281. *What is the comparative consistency of nerves in different periods of life?*

In early life the nerves are soft and pulpy. The brain itself is not in a condition to be applied to mental labors till about seven years of age. At the age of forty the nerves become smaller and dryer.

282. *When is the best time for study?*

In the morning, for then the brain is rested. The morning is also the best time for physical exercise. We should attend to the physical exercise first, and devote the remainder of morning hours to study. Physical health must be attended to, for if health fails all mental exercise to any extent is at an end. Exercising in the morning before commencing study will tend to preserve and invigorate health.

283. Is the nervous system easily diseased?

The main thing necessary for the general welfare of the nervous system is to attend to the general health of the body. If a person has nothing on which to exercise his nervous energy he is liable to disease. Employment of some kind is indispensable to the health of the nervous system. Long-continued trains of thought, however, are to the brain, what working one set of muscles incessantly all day, is to them: complete exhaustion. He that would last the longest, must occasionally turn his thoughts from his ordinary avocation completely, and give the brain rest. Every one, whether business man, student, farmer, or mechanic, needs a vacation once or twice a year, when, for a few days or weeks, he may break up the ordinary routine of life. To have a healthy condition of the nervous system, it is proper that the mind should have a variety of objects on which to dwell. Its efforts, however, should not be made spasmodically. There should be system; not working the brain to its utmost tension for a time, and then letting it lie idle, but working regularly and steadily.

## Chapter Seven.

**ORGANS OF THE EXTERNAL SENSES.**

284. What are the organs of external sense?

Those organs of the animal machine which bring it into relation with external objects are five: smell, sight, hearing, taste, and feeling, or sense of touch. The first four are situated in the head, while the organ of touch is distributed over the entire skin of the body.

**ORGAN OF SMELL.**

285. What is the organ of smell?

The nose is an organ admirably adapted to the office of smelling. The air, laden with odorous particles, can be drawn through it, and over the delicate membrane with which it is lined, near the surface of which commence numerous nerves, which unite with, and form the olfactory nerve, which carries its impression to the center of animal perception, the top of the medulla oblongata. There are four cavities to the nose, two through the upper jaw into the throat, by which the nose communicates with the lungs, and thus it is admirably adapted as a breathing organ, as well as smelling. The little hairs crossing the outer cavities of the nose are for the purpose of preventing the ingress of injurious particles of dust to the lungs. The nose is the natural passage for the external air to the lungs. It is by this sense that we are warned of the presence of decaying and unwholesome articles, and through it we experience a thousand delights from the fragrant odors of nature's flowers, &c. The odor of healthful food also quickens digestion.

286. *What is essential to a healthy conditon of this organ?*

The integrity of this organ requires that the mucous membrane lining the nose should be continually moistened and lubricated by its own exhalation and secretion. It is liable to become dry, so nature has provided it with facilities for abundant moisture. Colds, inflammation of the mucous surface, such narcotics as tobacco, snuff, smelling bottles of hartshorn, camphor, and all strong and pungent perfumery, weaken, paralyze, and sometimes utterly destroy, all perception of odors, and injuriously affect the whole brain through this sense. Sneezing has been said by some to be the voice of God in our nature commanding us to avoid what causes us to sneeze. Constant irritation of the mucous lining of the nose may in time hush this voice so that the peculiar sensibilities of the nose no longer warn us of intruders from that source.

## THE ORGAN OF SIGHT: THE EYE.

287. *What is the organ of sight?*

The eye; that wondrful organ whose healthful function enables us to see surrounding objects, and thus avoid many injuries, as well as experience many joys. The eye is of a globular form, composed of a number of humors, which are covered by membranes, and inclosed in several coats. On the front surface there is a slight depression, and in this is situated the crystalline lens. This is a body of considerable thickness and strength, and has the form of a double-convex lens. It is placed in a perpendicular direction immediately behind the pupil, and is kept in its situation by a membrane which is called its *capsule*.\* In front of the crystalline lens, and occupying the whole of the front part of the eye, is the aqueous humor. It is composed principally of water, with a few saline particles, and a very small portion of albumen. A curtain with an opening in its center floats in the aqueous humor, but is attached to one of the coats of the eye at its circumference. This curtain is called the iris, and the opening in it is the pupil. It derives its name from the various colors it has in different individuals, and it is the color of the iris that determines the color of the eye. All the light admitted into the eye passes through the pupil, which is dilated or contracted according to the intensity of the light and power of the eye. The eye has three coats. The outer or sclerotic coat—the white of the eye—is that to which the muscles that move the eye in various ways are attached. Within the sclerotic coat is the choroid coat, composed mostly of blood-vessels and nerves. The inner coat is called the retina. It is either an expansion of the optic nerve, or composed of nervous filaments attached to it.

288. How is the eye and its various parts moistened?

There are what are called lachrymal glands, which constantly supply the eyes with moisture, not only when they are open and in action, but also when closed and quiet in sleep. There are two small openings from the eyes into the nose,

---

\*The flattening of this lens by old age or other causes, produces the defect in vision known as long-sightedness. When more than ordinarily convex, it causes near-sightedness.

through which the fluid secreted by the lachrymal glands is conveyed. When these glands are much excited by irritations of the eyes or nose, or by strong emotion of the mind, they pour their fluid into the eyes faster than the small nasal ducts can convey it into the nose, and it flows down the cheeks in tears.

289. What is the medium of sight to the eye, and how is sight effected?

Light is the medium of vision, and the light conveys the impression of the object to the retina. A good illustration of the action of the eye may be made by cutting a hole in a window shutter large enough to receive a spectacle glass, excluding all light from the room except what comes through the hole. If the sun is shining brightly upon the shutter the rays of light will be seen in the room drawing together till they come to a focal point, and then the rays pass on diverging from one another, but the angle will be alike both sides of the focal point. At this focal point all the rays coming through the glass cross each other, so that the top rays at the glass are the bottom ones beyond the point. If a sheet of white paper be placed a little beyond the focal point, a beautiful miniature image will appear upon it of whatever the rays of light may come from, but this image will be upside down, and turned side for side, caused by the crossing of the rays of light. If instead of a spectacle glass, a small glass globe filled with water be placed in the hole in the window shutter, the rays will cross and diverge before they get through it, and the image will be thrown upon the back part of the globe. The interior of the eye is represented by the darkened room; the cornea by the transparent window glass; the iris by the shutter; the pupil by the hole through which the rays of light enter; the aqueous, crystalline, and vitreous humors, constitute a lens of so great a convexity, that the rays cross and diverge before they get through the globe, and throw their inverted image upon the retina. Here the mind perceives it, and by usage, views the object as though it were right side up.

290. What care is necessary in relation to the eye?

Reading early in the morning, before the just-wakened eyes are accustomed to the light, reading by twilight or lamp-light, are all injurious to the eyes. Never read facing the light of the lamp or window, but let the light shine over your shoulder upon your book. Those who read much should spend considerable time out-of-doors, looking at distant objects. Close application to reading tends to shorten the sight and weaken the eye. It is supposed one reason city people lose their eyesight sooner than those in the country, is because their sight is confined to objects near at hand. Another reason, however, is the impurity of the air, it being infected with smoke, coal-dust, &c. Smoke of any kind, especially tobacco smoke, is very injurious to the eyes. But it must ever be remembered that all organs of the body sympathize with each other and are affected by diseases of each other; so to cure diseases of the eyes there must be care given to the vital interests of the whole domain of organic life.

THE ORGAN OF HEARING: THE EAR.

291. What can you say of the construction of the ear?

This organ, which, with that of sight, ministers

to the intellectual and moral wants of man, as well as the physical, and relates us in duties, interests and pleasures to our fellows, exhibits in its structure a greater complexity than any other part of the human organization. The ear may be divided into the outer, the inner, and middle parts, and the auditory nerve. The outer part consists of the external ear, and the tube which leads to the membrane of the tympanum. The external ear is composed of cartilage, covered with a delicate skin, supplied with nerves and blood vessels. It inclines forward and is adapted to collect sounds, which it conveys through the tube. This tube is nearly an inch in length, made partly of cartilage and partly of bone. It has a number of small glands which secrete the wax. Its entrance is guarded by stiff hairs to prevent the ingress of foreign substances to the ear. The middle part of the organ embraces the tympanum and its membrane, the small bones of the ear, and the *eustachian tube*. This tube passes to the throat a little back of the palate. It is about two inches long, and largest at the throat. The membrane of the tympanum is placed at the bottom of the external tube. This membrane is placed obliquely, inclining downward and inward; it is tense, thin, and transparent. The tympanum, between the external and internal ear, is of irregular cylindrical form, with several openings. It contains the four little bones of the ear called the *hammer*, the *anvil*, the *round bone*, and the *stirrup*. Muscles of very small size move these bones in various directions. The internal ear is composed of three parts, and is situated in a part of the temporal bone near the base of the skull. Its parts are called the cochlea, the vestibule, and the semicircular canals. The first resembles the shell of a snail. The vestibule is a sort of porch or entry, which communicates with all the other parts. The three semicircular canals are all back of the cochlea and vestibule. The auditory nerve is distributed to the semicircular canals, the cochlea, and the vestibule, terminating in the form of a pulp.

292. What can you say of the action of this organ?

As the pupil of the eye contracts or dilates according to the amount of light transmitted to it, so the nerves of the ear act upon the muscles of the internal ear in proportion to the softness or harshness of the sound transmitted. The muscles move the chain of small bones so as to conduct the vibrations of sound across the tympanum to the internal ear. The contained air of the tympanum reverberates the sound, which is strengthened and modified by reflection from the walls, cells, and spaces of the ear. The impression of this sound is taken cognizance of by the auditory nerve. Of the peculiar action of this nerve, we shall have to content ourselves with admiring its wonderful operation without being able to solve the mystery as to *how* it acts. Of the philosophy of sound we can only say here that it is a vibration of the air caused in various ways, as by the striking of a bell, by singing of birds, or by the human speech.

293. What can you say in general terms of the ear?

It is one of the most useful of the organs of sense. It is attuned to the varied and sweet sounds of nature. Through it the persuasive tones of eloquence exert more power to stir or to stay the passions of man than all the arguments the ablest reasoner can present to the judgment. How important that we carefully guard and preserve this organ. It will be perceived that persons on becoming blind have a more acute sense of hearing than they had before. This hearing in some respects compensates for loss of sight.

294. Is the ear very liable to disease?

There are very few causes of derangement of the ear. Ear wax sometimes hardens in the ear. This can frequently be entirely relieved by several times dropping into the ear a few drops of pure sweet oil, and swabbing out the ear thoroughly with a little warm soft water and castile soap. Colds in the head, if they affect the hearing, must be very carefully avoided. In dullness of hearing, while you can hear distinctly the ticking of a watch placed against the side of the head, there is hope. Throat diseases, and scarlet fever, are very liable to leave children hard of hearing. This difficulty, as well as a permanent discharge from the ear, usually results from taking cold while recovering from the above diseases. The greatest care should be used that such results should not follow scarlet fever, &c. The difficulty may be outgrown, if not it is likely to grow worse as life advances.

## THE ORGAN OF TASTE.

295. What is the organ of taste?

The tongue. It is composed of muscular fibers, arranged in almost every direction. Between these muscles is a quantity of adipose substances. At the back part it is connected with the os hyoides by a muscular attachment. It is also attached to the epiglottis and lower jaw by the mucous membrane; this membrane forms a fold in front of the jaw and beneath the under surface of the tongue. The surface of the tongue is covered with four kinds of papillæ, supported by a dense layer of membrane. At the root of the tongue are a number of mucous glands. The tongue is abundantly supplied with blood by the lingual arteries. It has three nerves of large size: the *gustatory* branch of the fifth pair, the nerve of sensation and taste, distributed to the papillæ; the *glosso-pharyngeal*, to the mucous membrane, follicles, and glands. It is a nerve of sensation and motion; the *hypo-glossal* is the principal nerve of motion to the tongue, distributed to the muscles. The nerves of the sense of taste in the tongue terminate in the papillæ of the tongue, and are most numerous in the mucous membrane which covers the end of the tongue.

296. How is the sense of taste effected?

The papillæ on the surface of the tongue, when brought in contact with savory substances, are excited to that degree that they become erect and turgid, and convey to the appropriate nerves this sense. Here again is one of the wonders of the nervous system: how one nerve so nearly like another in its substance, can have the sense of taste, while the other may have the sense of smelling or hearing. It seems to be necessary in order for the sense of taste to be exercised, that the substance tasted should be soluble. There is also such a sympathetic relation existing between all the organs of the body, that their derangement affects measurably the organ of taste. When the nose is obstructed and injured, the sense of taste is affected.

297. What constitutes a healthy taste?

The integrity of the sense of taste enables us to

select, with accuracy, those alimentary substances just adapted to the wants of the nutritive apparatus. The sense of taste, like all the special senses, is highly educable, but is very generally depraved and perverted. Those persons who cannot realize any agreeable savor in any article of nutriment until the papillæ of the tongue are stung into action by salt, pepper, mustard, vinegar, or other pungents, have greatly blunted the sense of taste, and know but little of the real pleasures of eating. Such eat more to silence the goadings of a morbid appetite than to enjoy life. We should carefully avoid the use of every substance which blunts the use of taste: such as intoxicating liquors, tobacco, spices, salt, &c.

### THE ORGAN OF TOUCH.

298. Where is the sense of touch in the human body?

The nerves of feeling are the posterior roots of the spinal nerves, and some fibers of the fifth and eighth pairs of cerebral nerves. These nerves are distributed to the papillæ of the skin. These papillæ are small elevations on the surface of the body enclosing loops of blood-vessels and branches of sensatory nerves. It is not possible to puncture the body in any place with the finest needle without wounding both a blood-vessel and a nerve.

299. What is the structure of the skin?

The skin is composed of two layers, called the *derma* and *epiderma*. The *derma*, or true skin, is composed of elastic cellulo-fibrous tissue, abundantly supplied with blood-vessels, lymphatics and nerves. It is the color of this skin that gives the color of different races of men. The superficial strata of the derma is the papillæ spoken of above.

The epiderma, or cuticle, is the scarf skin which envelopes and protects the derma. Its internal surface is soft, its external surface is hard and horny. The pores of the epiderma are the openings of the perspiratory ducts, hair follicles, and glands. Of these there are supposed to be about seven millions on the surface of the body. The cuticle becomes very thick and hard on parts of the skin subject to much friction, as the bottoms of the feet, and insides of the hands.

300. Is the derma confined to the external surface of the body?

It is not. The same membrane lines the cavities of the mouth, nostrils, windpipe, air-passages, the cells of the lungs, the meat-pipe, stomach, intestinal tube, &c. The internal lining of the body is called the mucous membrane. The skin of the surface of the body, and that of the lungs and alimentary canal, in many respects resemble each other, especially in regard to the substances which they throw off from the system; and they are to a considerable extent reciprocal in their offices, the excess of one corresponding with the suppression of the other. Thus if the insensible perspiration of the external surface becomes checked by sudden exposure—by taking cold—the internal skin collects and disposes of this matter that would have passed from the surface of the body. The nerves of the internal skin connect with the nervous center of organic life, while the nerves of the external skin connect with the center of the nerves of animal life, the top of the medulla oblongata. Thus the external skin and internal mucous membrane sympathize in a powerful manner with each other. Irritations of the mucous membrane affect the external skin, and irritations and affections of the external skin also affect the mucous membrane.

301. *In what parts of the body is the sense of touch the most acute?*

The lips, tip of the tongue, and the inside of the last joints of the fingers. At these points the nerves are more numerous, and nearer the surface, and the outer skin is thinner, than at other points. The sense of touch may be educated and increased to a surprising degree. The blind are taught to read, and even to distinguish colors, by this touch. As to how the nerves take cognizance of hardness or softness of bodies, whether they are rough or smooth, hot or cold, is another wonder in the structure of the nervous system. It is by the degree of resistance required in the papillæ of the body when brought in contact with any substance, that it is supposed the mind forms its correct idea of their quality in these respects.

302. *What are called the appendages of the skin?*

The hair and nails. It is a fact, however, that each of these is dependent on an organism of nerves, vessels, &c., for its sustenance and production. The root of the hair, which is situated just beneath the skin, consists of a small oval pulp, invested by a sheath or capsule. That part of the hair in a state of growth is hollow, and filled with this pulp. The vigor of the hair depends on the vigor of its roots. The vigor and integrity of these roots depends on the general welfare of the body. Injury to the digestive organs, gluttony, intemperance, sensual excess of any kind, anger, grief, fear, &c., powerfully affect the roots of the hair, and thus the hair itself. Violent grief, or excessive fear, have whitened the hair, sometimes in a very few hours. The coloring matter is furnished by the bulb at the root of the hair, and the color of the hair is according to the color of the bulb. It is the unhealthy action of the root of the hair that causes its dry appearance, or its turning gray. All applications to the head, except those which give vigor to the roots of the hair, and healthiness to the skin of the head, are decidedly injurious. Dietetic errors, or abuse of the stomach, are of the greatest injury to the hair; so a proper regard to all the laws of our being is the only reasonable ground on which we can expect a healthy head of hair.

303. What can you say of the nails?

The nails have their roots and organs by which they are produced, yet they are themselves destitute of nerves and vessels. They do not sympathize so powerfully with the affections of the body and mind as the hair, but they are more or less moist and pliable, or dry and brittle, according to the general health of the body.

304. What offices are performed by the skin, aside from its sense of touch?

The skin, through its sweat ducts, acts as an eliminating organ, removing from the blood a large amount of impure matter. Copious sweating, as a general law, is debilitating to the body, as it exhausts the serum from the blood; this creates a thirst for water. This water is taken up by the absorbents, only to be immediately expelled again from the blood. So excessive drinking of even pure water, and sweating, causes both the absorbing and eliminating organs to do a great amount of unnecessary duty. The skin is also a breathing organ. In a vigorous state of the body, not too much confined by clothing, the action of the skin on the atmosphere is very much like that of the lungs. It absorbs oxygen, and throws off carbonic-acid gas. The amount of solid matter eliminated from the body through the skin daily is about 100 grains. Frequently exposing the entire surface of the naked body to the air of a well-lighted room, at the same time applying a slight friction to its surface by rubbing, is highly beneficial. The skin is also a universal regulator of the heat of the body. When the skin is in a vigorous and healthy condition it throws off the surplus heat, or retains the deficiency, according to the necessities of the body.

305. What is necessary to properly care for the skin and assist it in its functions?

*Bathing* is a great assistant to nature, in that it removes from the surface of the body effete matters that have been conveyed there through the pores of the skin. If these pores become closed, and the skin fails to throw off the matters of insensible perspiration, the lungs are oppressed, the head is giddy and painful, the mouth becomes parched and feverish, the heart troubled with palpitations, the kidneys irritated by excess of duty, the bowels become liable to gripings, spasms, exhausting diarrheas, or inflammatory attacks. It is then of the highest importance to keep the skin in a healthy condition.

306. What general rules should be followed in bathing?

With healthy persons a bath every other day, at a temperature congenial to their feelings, may be good. Soft water should invariably be used in bathing. A good tub to stand in and a good sponge are the only essential articles necessary to give a common bath. A soft towel or a cotton sheet should be used to wipe the body thoroughly dry on leaving the bath, after which the whole surface of the body should be rubbed with the bare hand till the skin feels soft and velvety, and a healthful glow is upon the surface of the body. In case of feeble persons, the labor of the bath should be performed by an attendant, they themselves remaining passive to prevent exhaustion of the body. Feeble persons should take a rest, or a nap, after a bath, before they exercise. After there is a thorough reaction from the bath, light gymnastics, walking, riding, or light labor in the open air, according to the strength of the individual, are beneficial. Persons in good health will not experience any difficulty in taking a general bath on first rising in the morning. For all, and especially the feeble, eleven o'clock in the forenoon is the best time for taking a bath. Never take a bath until at least two hours after a meal. Never take a bath when the body is in an exhausted condition. Swimming or bathing after performing a hard day's labor, is a very pernicious practice. Those who practice swimming are very liable to remain in the water too long.

As a general rule, water cool, but not cold enough to produce a chill, is best for persons in comparative health. Persons of low vitality should use tepid water, extremely feeble individuals should use warm water, cooling the bath before leaving it as their judgment shall dictate. Cold water we call 60°; cool, 60° to 72°; tepid, 72° to 85°; warm, 85° to 100°. Always, before taking a bath of any kind, the head should be wet in cool water, or a linen head-cap, of two thicknesses, wet in cool water, should be placed upon the head.

307. What is the cause of colds?

A large portion of the blood naturally flows through the superficial veins supplying the capillaries of the skin, which pour their exhalations of effete matter through the pores of the skin. When these pores become closed by exposure to sudden changes of temperature, the blood is thrown from the surface to the deep veins. In this case this effete matter accumulates on the mucous membrane of the internal organs, and causes a cold, irritation, inflammation, &c., varying in intensity according to the violence of the check in the circulation. People usually suppose a cold is always taken by passing from a warm to a colder atmosphere, but frequently passing from cold out-door air into a highly-heated room, will occasion a suppression of the external circulation, and produce a cold. The body when excessively cold should be warmed gradually. Colds are more frequently taken by unevenness of temperature, as for instance, having the room very warm, then letting the fire go down, then raising the temperature again, &c. Eating a full meal at night, after fasting all day, or eating to fullness or oppression, when the body is in a relaxed condition, produces the same change in the circulation.

308. What is the most effective way of curing a cold, fever, or any irritation caused by suppression of the external circulation?

The old plan would be to take a potion of physic. This is about on the plan of cutting off your finger to cure the head-ache. It may relieve it, but it does it by increasing, for a time, the inward irritation, and of course decreasing still more the strength of the system. In case the bowels need relief from mucous already collected, a tepid water injection is one of the mildest remedies. But, the real end to be gained, aside from this, is to open the pores and establish the natural circulation of the blood. A warm-water sweat, wet-sheet pack, a dripping sheet, &c., act directly on these pores. But avoid all harsh treatment to open the pores, such as the flesh brush, and crash-towel rubbing. These open the pores, it is true, but they leave them gaping wounds. They are thus not only in a condition to dispose of effete matter, but they permit the nutritive particles to pass off from the body through the capillaries, which have been exposed by this harsh treatment. Another, by no means slight, evil inflicted on the surface of the body, is in shaving the beard. Nature requires its growth. If you think you must shave, do it in cold soft water. Better still to keep the razor off your face.

309. What further care of the surface of the body is necessary?

It is highly important to give special attention to the clothing. It should always be warm in all seasons, as light and loose as possible without bodily discomfort. Cotton and linen are the best clothing for summer. Linen for under-clothes is best in hot weather. Flannel, next to the skin, is hurtful in all seasons. In wearing flannel, as a general rule, cotton or linen should be worn next to the skin. Fur neck-clothing and caps are bad; heating too much those parts of the body. Light-colored clothing is best for summer, because it repels heat. Females are apt to wear too great an amount of clothing about the back and hips. Garters, and tight waist-bands, are both injurious, hindering the circulation of the blood, and producing varicose veins and many other diseases. Every article worn during the day should be taken off the body and permitted to air during the night; and the night-clothes, and bed-clothing, should be well aired during the day. These should all be kept clean by frequent washing. The clothing should be so adjusted as in the greatest possible measure consistent with the proper temperature of the body, to admit of a free access of air to the whole surface, and of the most perfect freedom of circulation, respiration, and voluntary action. Regularity should always be observed also in clothing the body. Boots, shoes, hats, caps, thin and thick stockings, gloves, &c., when worn, should always be worn under similar circumstances, not indiscriminately changed or altered. If a part of the body usually protected by clothing be exposed to a current of cold air, the person will take cold sooner than to expose the whole body. Great care should be taken in clothing the limbs and arms, hands and feet properly. The clothing of females should be of such a length as to escape being wet and brought in contact with the tender ankles, and the feet and ankles should be protected from cold and wet. Rubbers are injurious, and should only be worn to protect the feet from wet.

## Chapter Eight.
### THE VISCERA.
*Figure XVI.*

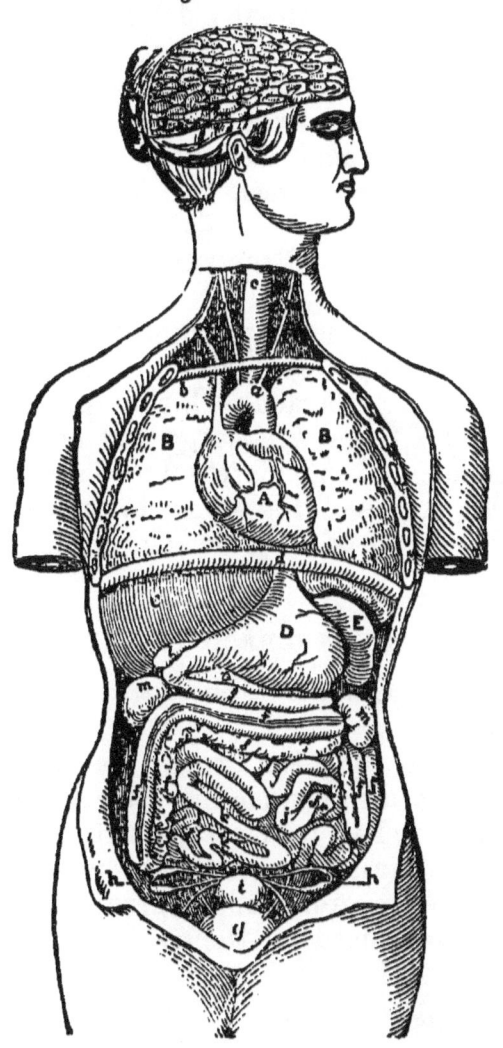

**VITAL SYSTEM.**

310. *What is Fig. XVI designed to represent?*

Those organs of the body called the viscera, which occupy the three great internal cavities of the system: the crano-spinal, thorax, and abdomen. The first is occupied by the brain and spinal marrow, which we have already described. The thoracic cavity, or chest, contains the heart and lungs. The abdominal cavity proper contains the stomach and intestines, liver, pancreas, spleen, kidneys, and supra-renal capsules. Its lower portion contains the intestines, the bladder, kidneys, &c.

The relative situation of the various parts may be seen by looking at *Fig. XVI.* A, the heart; B B, the lungs; $d$, is the diaphragm, which forms the partition between the thorax and abdomen; C, the liver; D, the stomach; E, the spleen; $m\ m$, the kidneys; $ff$ and $k\ k$, large and small intestines; $g$, bladder. Immediately under the letter $d$ is the cardiac orifice of the stomach, and at the right extremity or pit of the stomach, is the pyloric orifice.

### ORGANS OF THE THORACIC CAVITY.

311. *What are the organs, and their construction, in the thoracic cavity?*

The heart is one of these organs, but this we have already noticed in connection with the circulatory system, and shall have occasion to speak of it again after examining the action of the lungs. In the functions of respiration or breathing, and speaking, the organs used are the trachea or windpipe, lungs and diaphragm.

312. *What is the organ of voice?*

The structure composed of muscle and cartilage,

at the upper part of the wind-pipe, called the larynx, is the apparatus of voice; while the lungs and trachea are the organs of respiration.

313. What is the larynx, and where is it situated?

There is a funnel-shaped cavity back of the roots of the tongue, at the upper part of the wind-pipe, called the *pharynx*. This cavity is open from above, with canals coming from the nose, and the eustachian tubes from the ears; just in front of these is the pendulous body called the palate; in anatomy it is called the vail of the palate. In the act of swallowing, it is pressed back, closing the nasal canals and the eustachian tubes, so that nothing can pass into them. A little lower down, near the roots of the tongue, in the front part of the pharynx, opens the *larynx*, or the mouth of the wind-pipe. This is so situated that everything which is swallowed must pass directly over it. To prevent any of the food or drink from entering the wind-pipe, a small, oval-shaped valve of cartilage, called the epiglottis, is placed over it. This valve is always raised, except in the act of swallowing, when it shuts down over the orifice, and completely closes it for an instant, while the food or other substances are passing over it; then it immediately opens, that breathing may not be interfered with. It is in the *larynx* that all the modifications of the voice are produced, by the air passing through it from the lungs.

314. What is the construction of the larynx?

It is a short tube, of an hour-glass form, composed of cartilages, ligaments, muscles, vessels, nerves, and mucous membrane. The larynx is composed of five cartilages; the first of these produces, at the upper part of the neck, the prominence called *Adam's apple*. The larynx has twelve ligaments, or vocal cords, attached in front to the receding angle of the thyroid cartilage, and extending backward. It has eight muscles; these are used in opening and closing the glottis, and in regulating the position and tension of the vocal cords. The mucous membrane of the larynx is the same as that of the mouth, which is prolonged through it into the bronchial tubes and the lungs. The arteries are the superior and inferior thyroid. The nerves are branches of the pneumogastric.

315. How are the different tones of the human voice produced?

The voice is formed in the larynx, and all its modifications are produced by the simple expulsion of air from the lungs, when the vocal ligaments or cords are held in a certain degree of tension. The sound is occasioned by the vibration of the vocal ligaments. Speech is a modification of voice-sounds in the cavity of the mouth. The articulating organs are the tongue, palate, lips and teeth. The cavities of the nose also modify the speech. The English language may be reduced to forty-four rudimental sounds, or elements, sixteen of which are vowels, and twenty-eight are consonants. The muscles which stretch or relax the vocal ligaments are alone concerned in the voice. The *pitch* of the tones is regulated by the tension of the vocal cords. The *volume* or intensity depends on the capacity of the lungs, length of the trachea, or wind-pipe, and the force with which the air is expelled, and the flexibility of the vocal cords. In the male the vocal cords are longer than in the female, in the proportion of three to two, which renders the male voice usually an octave lower. The natural compass of voice, in most persons, is two octaves, or twenty-four semitones. Singers are capable of producing ten distinct intervals between each semitone, making 240 intervals, requiring as many different states of tension of the vocal cords, all of which are producible at pleasure, and without a greater variation of the length of the cords than one-fifth of an inch. One of the most wonderful feats accomplished in the human body, is the precision with which the will determines the exact degree of tension necessary to produce a given note in an instant of time, after the mind has decided the note required. How sad to think that an instrument so nicely and wonderfully constructed as the human voice, should ever be used in defaming its Maker, or harshly speaking to our fellow creatures.

316. What is necessary in training the human voice?

If speech is defective, the precise cause must be noticed, and the difficulty be removed, or overcome by exercise. If persons stammer, induce them to speak with the mouth open, and with the lungs filled with air. If they lisp, reading and speaking with the teeth closed will help it. Reading aloud, shouting, singing, and laughing, are healthful exercises, promoting digestion, and giving action to the lungs and abdominal muscles.

## THE TRACHEA, OR WIND-PIPE.

317. What is the trachea?

It is the *wind-pipe*, which extends from the larynx down to the third dorsal vertebra, where it divides into the right and left branch. The right passes off to the upper part of the right lung at nearly right angles; the left is smaller, and descends obliquely beneath the arch of the aorta, to the left lung. It is kept in a distended form by twenty-four cartilaginous rings, connected with each other by a membranous texture. These rings are not entire circles, but about one-third of the circle, and that on the back side, directly in front of the œsophagus or meat pipe, is occupied by a membranous texture of muscular fibers running in the direction of the rings, so that their contraction decreases the caliber of the wind-pipe. When the food is descending the meat-pipe this muscular portion yields, so that the passage of the food is not obstructed, as would be the case if the rings passed entirely around. As the branches of the wind-pipe become subdivided in the substance of the lungs, these rings become softened down, and gradually disappear, leaving nothing but the membranous forms of the air tubes.

### THE LUNGS.

318. What is the structure of the lungs?

They are two conical-shaped organs, occupying the cavity of the chest on each side of the heart, from which they are separated by a membranous partition, the mediastinum. Their color is pinkish gray, marked with black. Each lung is divided into two lobes by a long, deep fissure. In the right lung the upper lobe is subdivided by a second fissure. The air cells in each lobe communicate with each other, but not with those of another lobe. The lungs rest on the convex surface of the diaphragm. The *root* of each lung comprises the pulmonary artery and veins, and bronchial tubes, with the bronchial vessels and pulmonary plexuses of nerves. They are comprised of ramifications of the bronchial tubes, terminating in intercellular passages and air cells. It is supposed that there are not less than one hundred million air cells in the lungs. The mucous membrane of the lungs presents an extent of surface of twenty-one thousand square inches; supposed to be greater than the entire surface of the skin of the body. All the air tubes, vessels, and nerves of the lungs are closely knit together into one general texture, by a delicate cellular tissue, and the whole mass, on each side, is enveloped in the serous membrane as an external coat. The pulmonary artery, which transmits the venous blood from the heart to the lungs, terminates in a minute network of capillary vessels, distributed through the walls of the air-passages and air-cells; these converge to form the pulmonary veins, which return the arterial blood to the heart.

319. What is the pleura?

It is the serous membrane which lines the thoracic cavity, and divides it into two chambers, by passing double across it from the breast-bone to the back-bone, thus forming a closed sack for each lung, and embracing the heart, the large blood-vessels and the meat-pipe, between the two sheets of the middle partition. If the two lungs occupied only one cavity, then any perforation of the walls of that cavity, by disease or otherwise, so that the external air could rush into it, would at once arrest the function of respiration, and immediate death would result. But now, if by any means one lung is disabled, it can lie still while the other continues faithfully to perform its functions. Such is the wisdom and goodness of our great Creator.

320. How is the process of breathing accomplished by the lungs?

The movement of the lungs is partly voluntary, and partly involuntary. Voluntary for the purpose of being guided by the will, and assisting speech; involuntary, that they may be acted upon by the involuntary nerves and muscles of organic life. The atmosphere presses on the surface of the body at the rate of fifteen pounds to every square inch. This pressure being the same on all parts of the body, we do not feel it. When the diaphragm is drawn down, and the breast-bone and ribs elevated, the cavity of the chest is much enlarged, and the external air rushes into the air cells, distending them in proportion to the dilation of the thorax, and keeping the surface of the lungs all the while accurately in contact with the walls of the chest in all their movements. Immediately after an inhalation of breath, all the muscles employed in expanding the cavity of the chest contract, and the ribs and diaphragm return to their natural position. By these means, and by the contraction of the muscles of the air tubes, the air is expelled from the lungs. It will be readily seen that, in the function of breathing, the lungs themselves are entirely passive, and that in order to breathe freely there should be ample room to expand the chest. When the ribs are confined by tight clothing the diaphragm is compelled to carry on the function of breathing alone, but in this case respiration is much restrained.

321. What further can you say of the structure of the lungs?

The sides or walls of air cells are formed of a very thin, transparent membrane, and the capillary vessels are placed between the walls of two

# THE VISCERA. 151

adjacent cells, so as to be exposed to the action of the air on both sides. It is calculated that about 266 cubic feet of air pass through the lungs of a medium-sized man in twenty-four hours. It is calculated that nearly twelve pints are ordinarily present in the lungs at one time, and that about a pint is inhaled and exhaled at one inspiration, so it would require twelve ordinary breaths to displace the air in the lungs, and supply entirely new air. This of course has reference to involuntary breathing, and not to deep and full inspirations controlled by the will.

*Figure XVII.*

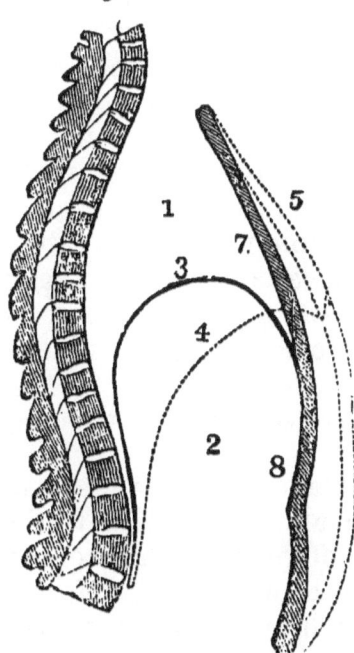

322. What is *Fig. XVII* designed to illustrate?

A side view of the chest and abdomen in respiration. 1, cavity of the chest; 2, cavity of the abdomen; 3, line of direction for the diaphragm when relaxed in expiration; 4, line of direction when contracted in inspiration; 5, 6, position of the front walls of the chest and abdomen in inspiration; 7, 8, their position in expiration.

ACTION OF THE DIAPHRAGM.

323. What is accomplished by the inhaling of air into the lungs?

The air is digested, and the lungs form from it a principle convertible into the substance of the blood. They also constantly receive from the air, and transmit to the blood, a replenishing stream of that electric, magnetic, vital property, on which the nervous influence depends. The composition of one hundred parts of air is, twenty parts of oxygen, seventy-nine of nitrogen, one of carbonic-acid gas. Other gaseous matters are sometimes found in the air, but are not natural constituents. The air expired from the lungs has lost about sixteen parts of its oxygen, and gained about fourteen parts of carbon. This decarbonization and oxygenation of the blood changes it from a dark purple to a bright, florid color. The oxygen does not go into the blood as oxygen, but is converted into electricity, the same as food in its digestion is converted into chyle. The variation in the quantity of nitrogen taken up from the air in the lungs, must be regulated by the excess or deficiency of the nitrogenous principle in the food eaten. It is supposed that thirty-seven ounces of oxygen are retained from the air, and fourteen ounces of carbon thrown off by the blood through the lungs every twenty-four hours.

324. *What is essential in reference to the air, that it may impart vigor and length of days?*

It is of the utmost importance to the welfare of the body, that pure air should be supplied to the lungs at every inspiration of breath. The organ of smell is a grand sentinel to warn us of infected and impure air; happy is he who listens to the dictates of nature in this respect. It is impossible for the lungs to be fully expanded in an impure atmosphere. It irritates the lungs, and the air cells spasmodically contract to keep out such

## The Viscera. 153

air. It is no wonder that those who live in the smoke and din of cities, and inhale the impurities of a hundred cess pools, and neglect to exercise their lungs by expanding them to their utmost capacity, many times a day, in pure air, should find their lungs diminishing in size, and pulmonary consumption marking them as its victims.

*Figure XVIII.*

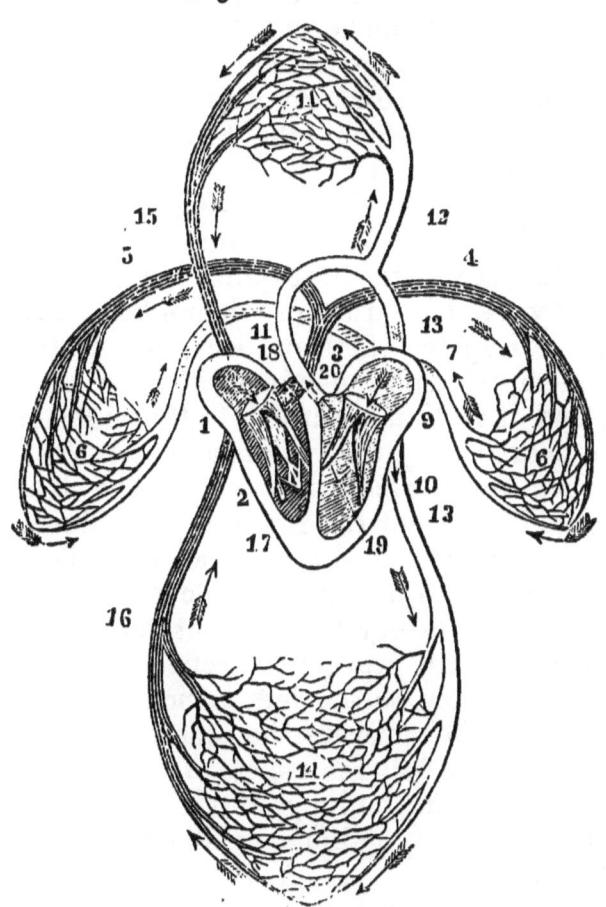

**PURIFICATION OF THE BLOOD.**

325. What is *Fig. XVIII* designed to illustrate?

The circuit of the blood in the system, and its course into the lungs, where its purification is accomplished. From the right ventricle of the heart (2), the dark, impure blood is forced into the pulmonary artery (3), and its branches (4, 5), carry the blood to the left and right lung. In the capillary vessels (6, 6) of the lungs, the blood becomes pure, or of a red color, and is returned to the left auricle of the heart (9) by the veins (7, 8). From the left auricle the pure blood passes into the left ventricle (10). By a forcible contraction of the left ventricle of the heart, the blood is thrown into the aorta (11). Its branches (12, 13, 13,) carry the pure blood to every organ or part of the body. The divisions and sub-divisions of the aorta terminate in capillary vessels, represented by 14, 14. In these hair-like vessels the blood takes up the effete and worn-out particles of the system, which render it dark colored and impure, and it is thus returned to the right auricle of the heart (1) by the *vena cava descendens* (15), and *vena cava ascendens* (16). The tricuspid valves (17) prevent the reflow of the blood from the right ventricle to the right auricle. The semilunar valves (18) prevent the blood passing from the pulmonary artery to the right ventricle. The mitral valves (19) prevent the reflow of blood from the left ventricle to the left auricle. The semilunar valves (20) prevent the reflow of blood from the aorta to the left ventricle. The amount of air required daily in the lungs is three thousand two hundred and forty gallons, or about eight barrels per hour.

326. What do we find, then, most essential to keep the lungs in a healthy condition?

To emancipate them entirely from compression, so that with the greatest ease they can be filled to their utmost capacity. Just in the same ratio that the lungs are confined, and the breathing capacity of the lungs decreased, is the life shortened. It is a good exercise, especially for students, in-door mechanics, and all persons of sedentary habits, several times in a day, to fill their lungs to their utmost capacity half a dozen or a dozen times in succession, at the same time whirling their arms, or swinging them back behind them. Such exercise should be commenced carefully.

327. Why is air which has been breathed considered poisonous?

When the oxygen has been exhausted from the air by breathing, a person will soon expire in the air. This air is poison merely because it has no vitalizing element in it, nothing to sustain life. A fire or lamp will go out where this oxygen is consumed. In wells, or deep vaults, a lighted lamp should be lowered before a person descends into them. If the lamp is extinguished, a person could not live in the place. Carbonic-acid gas is heavier than common air, and settles to the bottom of the room or vault. In closely-crowded, illy-ventilated rooms, the purest air is in the middle of the room. A hundred persons confined in a room thirty feet in length and breadth, and the usual height, would render the whole air unfit to breathe in less than two hours. An ordinary lamp or gas-burner consumes as much air as four persons; a common stove consumes more air than fifty persons. In addition to this, a large amount of impurity is communicated to the air by the exhalations from the skin. So it will be readily seen, that the air of meeting places, crowded factories and work-shops, is in a few minutes unfit to sustain life. How important, then, that such places be well ventilated.

328. What is essential to good ventilation?

That the air should be pure. A lamp burning in a sick room, or a smoking lamp, requires special care that pure air is supplied to the room; but always remove, as soon as possible, the smoky lamp from the house. Light is essential to thorough ventilation. Rays of light, especially sun-light, have a wonderful, invigorating influence on the air. Those living in low cellars, or damp, shady sides of streets, or damp, dark places in cities, cannot expect to enjoy perfect health. In prevailing epidemics, as cholera, it prevails the worst in dark and damp places, and the shady sides of streets. The rooms mostly occupied in dwelling houses, should be so arranged as to have plenty of light admitted to them, and so that pure air can pass through them. It is a decided benefit, in a room where fire is burning, to keep constantly a dish of water on the stove, that the air may be moistened by the steam. It is very unhealthy to sleep in rooms in which are several house plants, or which are surrounded by, or immediately adjacent to, dense foliage. Vegetation absorbs carbon during the day, and throws off oxygen, but in the night this process is reversed, and it absorbs oxygen and throws off carbonic-acid gas. In rooms difficult of ventilation, by swinging the door back and forth rapidly several times in succession, the impure air may be pumped out of it, and pure air will rush in to fill its place.

329. What is the disease to which the lungs are most liable?

Inflammation of the lungs, called in its various stages and manifestations, pneumonia, pneumonitis, lung fever, pleurisy, &c. The main cause producing it is exposing the body to extremes of temperature, unequal exposure of the body, cold or wet feet, exposing the body to cold or wet, when it is exhausted either by over-exertion, or loss of sleep. The tepid wet-sheet pack is good in such cases to allay the fever on the surface of the body. Pain or oppression in breathing, or soreness in the lungs, may be relieved by the compress, made of a linen cloth of two or more thicknesses, wet and placed over the lungs, with sufficient dry flannel over it to occasion a slight perspiration. This should be wet several times a day. In a violent attack of lung complaint, if possible, immediately get the aid of some one of experience in the use of water. But it is better to studiously avoid all those causes which are liable to induce diseases of the lungs. Keep your feet warm and dry, your head cool, get a proper amount of rest every day, and labor within your strength, always avoiding exhaustion of vital power as much as possible. The thirst which attends a fever, is a demand of nature. This demand should be supplied by freely drinking of pure, soft water. The bowels should be kept regular by injections of tepid water.

## THE ABDOMINAL VISCERA.

330. What is the abdominal viscera?

As before stated, it is the organs of the abdominal cavity; comprising the stomach, alimentary canal, liver, pancreas, spleen, and kidneys, with the supra-renal capsules. For convenience, the abdominal cavity is divided into three zones, called the upper, middle, and lower zone. In the upper zone is found the liver, extending from the right to the left side; the stomach and spleen on the left, and the pancreas and duodenum behind. This zone extends from the diaphragm to the lower front point of the ribs. In the middle zone is found the upper part of the ascending and descending colon, omentum, small intestines, and mesentery, and behind, the kidneys and supra-renal capsules. In the lower zone is the inferior portion of the omentum and small intestines, lower portion of the ascending and descending colon, bladder, ureters, &c.

331. What lines the abdominal cavity?

The peritoneum is the serous membrane of the abdominal cavity; it invests each organ separately, and is then reflected upon the surrounding ones, enclosing the whole in a sac. The diaphragm is lined by two layers, which pass to the upper surface of the liver. They form its coronary and lateral ligaments. Passing around the liver, they meet on its under surface and pass to the stomach, forming the lesser omentum. These layers then surround the stomach, and, descending in front of the intestines, form the great omentum. They then surround the transverse colon, and passing backward to the spine, form the meso-colon, where the layers separate. The back layer ascends in front of the pancreas and aorta to the diaphragm. The front descends, and, after investing all the small intestines, it returns to the spine, thus forming the mesentery. Descending into the pelvis, it forms the meso-rectum, and a pouch called the recto-vesical fold, between the rectum and bladder. It then ascends upon the neck of the bladder, forming its false ligaments, and returns upon the front walls of the abdomen to the diaphragm.

## THE VISCERA.

### Figure XIX.

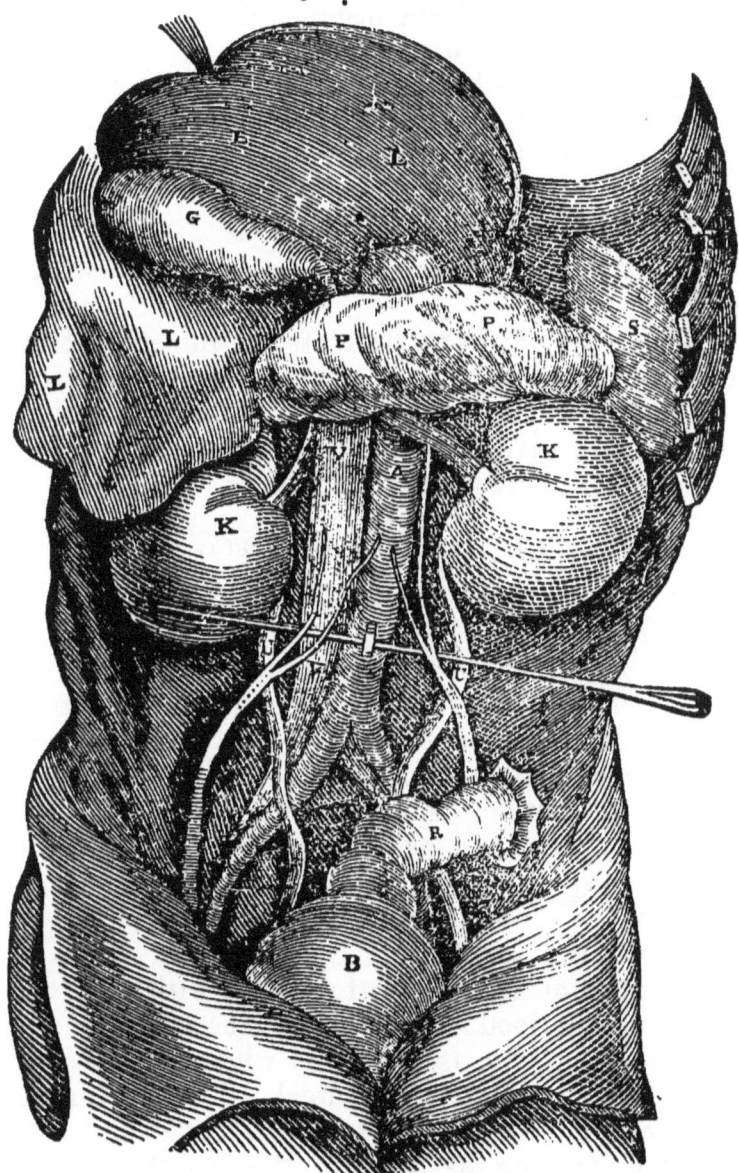

ORGANS OF THE ABDOMEN.

332. What does *Fig. XIX* illustrate?

The position of the organs of the abdominal cavity. A is the descending aorta: B, the bladder; G, the gall bladder; K, the kidneys; L, the liver turned up, showing its under surface; P, the pancreas; R, the rectum; S, the spleen; N, the ureters; V, the vena cava. In this figure the intestines are mostly remòved.

333. What are the offices of the great omentum and messentery?

The great omentum protects the intestines from cold and friction, and facilitates their movements. The mesentery retains the small intestines in their places, and gives passage to the mesenteric arteries, veins, nerves, and lymphatics.

### THE ABDOMINAL CAVITY.

334. What is the alimentary canal?

It is a continuous tube from the mouth to the anus. It is distributed into various portions, named as follows: The mouth, pharynx, œsophagus, stomach, and intestines. The intestines are sub-divided into the small, which are distinguished as the duodenum, jejunum, and ileum; and the large, distinguished as æcum, colon, and rectum.

335. What office does the mouth hold in the work of alimentation?

It is in the mouth that the food is prepared to enter the stomach. It is furnished with teeth for grinding the food. Time should be taken in eating the food, to mash it finely with the teeth, for this greatly lightens the work of the stomach. The mouth, and not the stomach, is the mill; so let the grinding be done in the mouth. There are also connected with the mouth salivary glands.

Of these there are three, called the parotid, sub-maxillary, and sub-lingual. The first of these, which is the largest, is situated just in front of the external ear. Its ducts communicate with the mouth near the second double tooth. The second pair of these glands lie just within the lower edge of the under jaw, on each side. The third and smallest pair lie under the roots of the tongue, uniting on the middle line of the tongue.

336. What is the office of the salivary glands?

They secrete the saliva or solvent fluid of the mouth, and pour it freely into the oral cavity during the process of mastication, and whenever any exciting substance is taken into the mouth. The smell, and sight, and even the thoughts, of savory or disgusting substances, will cause an increased secretion and flow of saliva. By the action of the saliva upon the food in the mouth, the food is not only prepared for swallowing, but the process of digestion is commenced. This work is performed to a greater or less extent according to the length of time the food is kept in the mouth. The process of chewing increases the flow of saliva. If the food is not properly masticated, the stomach is irritated, and *all* irritations of the stomach greatly affect the condition of the salivary glands, and the nature of their secretion.

337. What injury is done to the stomach by improper mastication?

A four-fold injury is done to the stomach: 1, It compels the stomach to receive the food more rapidly than is consistent with its welfare. 2, It compels the stomach to secrete a larger quantity of solvent fluid than would be necessary if the functions of the mouth had been properly performed.

3, It compels the stomach, at great inconvenience, to reduce by maceration those masses which ought to have been broken down and finely ground by the teeth. 4, By increasing the duration and difficulty of gastric digestion, it increases the expenditure of the functional powers of the stomach, and thus causes a greater degree of vital exhaustion in that organ, tending to debility and disease.

338. What are the properties of the gastric juice?

The gastric juice is supposed to contain muriatic and acetic acids, phosphates and muriates of potassa, soda, magnesia, and lime. This composition, however, differs with the quality or kinds of food taken. The active principle of gastric juice is called pepsin, which has the power of exciting chemical changes in the particles of other substances without itself decomposing. But no pepsin prepared by a chemist can accomplish the effect produced in nature's great laboratory.

339. Does drinking with our food injure the action of the salivary glands?

To drink cool, soft water while eating, if we are thirsty, is probably not injurious, if the mouth is cleared from food before we drink. Washing down our food with drink of any kind is injurious. The food should be moistened sufficiently to swallow with the saliva, for reasons before assigned.

340. What are the tonsils, and what is their use?

They are two almond-shaped bodies, situated in each side of the back part of the mouth. They are clusters of mucous glands. They pour out mucous and lubricate the food forced between them as it is being swallowed. The tonsils are liable to swell and become troublesome, being sore to the touch externally, and partly closing the passages within; in which case it is sometimes advisable to have them cut out by a skillful surgeon.

341. What is the œsophagus?

The œsophagus is the continuation of the alimentary canal from the pharynx to the stomach. It is sometimes called the meat-pipe. It is some twelve or fifteen inches in length. In its descending course along the spine, it inclines to the left in the neck, to the right in the upper part of the thorax, and to the left again as it passes through the back part of the diaphragm. It terminates at the stomach, and serves the office of conveying the food from the pharynx to the stomach.

342. How is the food swallowed?

When the food is prepared for swallowing, it is gathered back upon the arch of the tongue, whence it is suddenly launched into the pharynx, and passes through the meat-pipe to the stomach. The food, it will be remembered, in passing to the meat-pipe, passes over the glottis or top of the wind-pipe, as well as past the nasal cavities and eustachian tubes. These are all closed in the act of swallowing, that the food or drink may not pass into them. At the instant the food is launched from the arch of the tongue, the muscles of the pharynx contract, shortening the pharynx, and raising up the larynx; at the same instant the vail of the palate is pressed back, and closes the nasal canals and the tubes coming from the ears; the epiglottis shuts down and closes the glottis, and the pharynx darts up, and seizes the descending mass, and suddenly dropping down, presses it into the meat-pipe. As soon as the œsophagus receives the food, its muscular coat contracts upon it from above downward, and presses it onward into the stomach; and at the same time the mucous follicles, situated in this narrow passage, pour out their lubricating fluid to shield the nerves and vessels of the lining membrane, and to facilitate the movement of the descending mass. When the food has passed from the œsophagus, its lower portion continues to contract upon the stomach to prevent the food passing back during the action of the stomach.

343. What is the stomach?

The stomach is an expansion of the alimentary tube, its greater or splenic end being brought in contact with the concave surface of the spleen. The lesser or pyloric end extends into the epigastric region. Its opening into the œsophagus is from its upper side, and on account of its proximity to the heart, it is called the cardiac orifice. The inferior mouth of the stomach, which opens into the small intestine, is but a little lower than that at which the food enters. This opening from the stomach is called the pyloric orifice. The stomach is ordinarily capable of containing from one to two quarts. It may be enlarged by gluttony, or diminished in size by disease.

344. How does the stomach act upon the food?

As stated in the chapter on the muscular system, the stomach has muscles extending both ways; around it and lengthwise. When the food is felt by the nerves in the stomach, the muscular fibers are called into a rapid and vigorous action, the whole stomach is thrown into a gentle commotion, by which the food is carried around the gastric cavity, and everywhere pressed against the internal surface. This excites the glands, which secrete a thin, transparent fluid called gastric juice, which very soon begins, like sensible perspiration, to exude from the mucous membrane, in small drops, and mingle with the food. After the first portion of food has been carried about the gastric cavity, and freely mixed with this fluid, if the stomach be not crowded too fast with food, its muscles relax to some extent, and it is prepared to receive another portion of food, which undergoes the same process as the first. These operations are continued, until the stomach is distended with food, and the meal is finished. Then the muscular action is less rapid; a gentle, undulating motion takes place, and is kept up, till the function of the stomach is completed, and its contents poured through the pyloric orifice into the small intestine.

345. What is the pyloric orifice?

It is a thick band of muscular fibers, forming a powerful ring, which, together with a thickening or folding of the mucous membrane upon itself, forms what is called the valve of pylorus, or, more commonly called, the pylorus or "gate-keeper." When this ring is contracted, the orifice is closed. Its office is to prevent the contents of the stomach from passing into the small intestine in a crude and undigested state; but when a portion of food has become prepared to pass from the stomach, it is carried along by the muscular action of the stomach to the pylorus, which, by a peculiar organic instinct, perceives its character and condition, and immediately opens and suffers it to pass into that portion of the small intestine called the duodenum. When the pylorus is in a healthy condition, if a crude mass of undigested food attempts to pass into the duodenum with the chyme, it immediately closes, forcing the matter back, to be subjected again to the action of the stomach. If the substance, after a few such efforts to pass, is discovered to be of an indigestible nature, the orifice either opens and allows it to pass, or by a convulsive effort the muscles of the stomach contract upon it, ejecting it through the meat-pipe and mouth.

346. What change is effected upon the food in the stomach, and how?

The nutritious portion of the food, or that which can be assimilated and elaborated for the building-up processes of the system, is converted into a substance called chyme. This substance differs from anything in the food when it is received into the stomach. It is identical in character, whatever may be the food from which it is formed; but it differs in quality with the quality of the food eaten. This change in the food in the stomach is principally effected by the vital action of the gastric juice. The chemical and physiological character of the gastric juice is very considerably affected by the dietetic habits, general state of the health, the affections of the mind, and the condition of the stomach. After the food is received into the stomach, and the process of gastric digestion is to be commenced, the temperature of the stomach is raised to about 100 or 104° Farenheit, and this gastric juice, when the stomach is heated, is of a solvent nature, having power to chymify the food.

347. Is all the food in the stomach chymified at once?

No; that portion of the food which comes in contact with the mucous membrane is converted into chyme, and by the muscular action is forced through the pylorus. Then another portion of the food comes in contact with it, and so on until all that can be is chymified. The innutritious portion is separated from the nutritious, and reduced to such a condition as to fit it to pass along the alimentary tube as excrementitious matter. As soon as the food is chymified and passes from the stomach, that organ is left entirely empty and clean, and contracts upon itself, and remains in this state till some alimentary or other substance is introduced into it. When the stomach is in a healthy condition, after it has had a suitable amount of rest, and just when nature wants it, it causes the manifestation of feeling called hunger. Some mistake a tired feeling in the stomach for hunger. It is folly to talk of real hunger in less than six hours from the time we have taken a full meal.

348. What time is requisite for the stomach to digest a meal?

This varies with the quality of the food, the condition of the stomach, and the varying circumstances and condition of the individual; but as a general rule, the food received at an ordinary meal undergoes the process of gastric digestion, and passes from the stomach into the duodenum, in from two to five hours. When water, or milk, or liquid food holding in solution particles of aliment, such as soups, &c., are taken into the stomach, the water is all taken up by the radicles of the veins, and carried unchanged into the circulation, before the process of digestion is commenced. If the stomach is in a healthy condition, the water is all removed by absorption in a very few minutes. In some cases of chronic disease, as dyspepsia, the water remains in the stomach, causing flatulence and acidity, and retarding digestion for hours, until large portions of it are thrown up with portions of undigested food.

849. Why is it essential to keep the stomach in a healthy condition?

Because the alimentary cavity is the principal avenue through which the causes of disease commit their depredations on the vital domain. The stomach is peculiarly a center of irritation, and a starting-point of disease to the whole body. Whatever is unfriendly to the vital interests, that impairs the nervous power, or muscular contractility, or in any way disturbs the functions of the stomach, more or less impairs the quality of the chyme elaborated from the food, and this directly leads to a deterioration of all the fluids and solids of the body. Let it also be borne in mind that the heart, lungs, liver, and all other organs of the body, directly sympathize with all irritations and disease of the stomach?

350. How must the stomach act to be kept constantly in a healthy condition?

All its mucous surface needs to be in a condition to be brought in contact with the alimentary substance placed in it; but this cannot be the case when it is seared over with pepper, spices, and other condiments. Articles of an injurious character, and causing feelings of repugnance in the stomach when first placed in it, irritate it, and destroy its organic sensibility in a great measure, or at least to that extent that substances of the most deleterious character may be thrown into the gastric cavity, working out the destruction of our lives, and we not be conscious of it. The appetite, and even the stomach, may be so depraved that they will receive these life-destroying substances with great satisfaction, and the person using them declare that they are not injurious, because they sit well upon their stomachs; when the facts are, that the stomach has lost the power to discriminate between good and evil, and to give the proper alarm when the vital interests are in jeopardy.

351. What is connected with the pyloric orifice of the stomach?

The alimentary tube or small intestine, which is six or eight times the length of the body, and is nicely folded so as to be brought into a small compass. A portion of this is seen at *f, f, k, Fig. XVI.* It is about twenty-five feet in length, and is divided into three parts, called by physiologists the duodenum, the jejunum, and the ileum. This tube at its lower portion suddenly expands into what is called the colon, which is more capacious than the small intestine. The colon ascends to the stomach on the right side, arches over the whole volume of the small intestine, and descends on the left side, forming, in its lowest part, what is called the sigmoid flexure, which is in the shape of a letter S. This enters into the formation of a smaller tube called the rectum, at the lower end of which the mucous membrane again blends with the outer skin of the body.

352. How is the intestinal tube lubricated so as to preserve it from the injurious action of substances introduced into it?

Throughout its whole extent it is furnished with glandular follicles, which copiously secrete, and pour upon its surface, a lubricating and sheathing mucous. By this mucous the whole extent of the tube is preserved from injury.

353. What is the second stomach?

It is the first portion of the alimentary tube below the stomach, called the duodenum; which, as its name signifies, is about twelve finger-breadths long. From the pyloric orifice of the stomach it extends upward and backward toward the liver, turns down, and then, deeply situated, (see *f*, *Fig. XVI*,) crosses to the left side and comes forward again.

354. What are the appendages of the duodenum?

Its glands, called Bruner's, Lieberkun's, and Peyer's, and the pancreas, the liver, and the gall bladder.

355. What is the pancreas?

The pancreas is a long, flat gland, six inches long, one inch thick, and weighs from four to six ounces; situated across the abdomen, behind the stomach, opposite the first and second lumbar vertebræ. Its greater end or head is toward the right, surrounded by the curve of the duodenum; the lesser end extends to the left as far as the spleen. The duct through which the pancreas pours its juice into the duodenum, enters the duodenum about four inches below the pyloric orifice of the stomach. The pancreas, in structure and in the character of its secretion, very closely resembles the salivary glands.

356. What is the office of the pancreas?

To secrete and pour into the second stomach the pancreatic fluid, which is employed in perfecting the process of chymification in the small intestine.

357. What other appendage to the duodenum has its duct, entering at the same point with the pancreatic duct?

The largest gland in the body, called the liver.

It is situated at the top of the abdominal cavity, and lies immediately under the diaphragm, and mostly on the right side. It measures about twelve inches through its longest diameter, and weighs about four pounds. It is divided into a large lobe and two small ones. On the lower surface of the large lobe, which lies on the right side, is formed a membranous reservoir, called the gall-bladder, which is also lined by the mucous membrane. The common biliary duct, after proceeding a short distance from the small intestine, gives off a tube called the cystic duct, which goes to the gall-bladder. The capacity of the gall-bladder is from one to three ounces. The remaining portion of the tube is called the hepatic duct, which soon divides, forming two tubes, one of which goes to the right and the other to the left lobe of the liver. The liver is held in its place, and attached to the diaphragm, by five ligaments. The nerves of the liver are very numerous, and by them it is brought into powerful sympathetic relations with the stomach. The gall-bladder serves simply as a reservoir for the bile.

858. What is the structure of the liver?

It is merely a collection of parts similar to each other, called lobules; these are bound together by a small number of sinewy fibers.

359. What is the office of the liver?

To secrete the bile of the venous blood from the capillaries which penetrate every part of the liver, and pour it into the bile ducts. The grand function of the liver is that of a cleansing organ, and is evidently designed to act the part of a *filter*, in separating impurities from the venous blood of the portal system, coming from the tissues of the alimentary canal. If the liver does not duly eliminate the bile, the blood will become thick, the skin dingy, the head oppressed, the mind confused, the nerves weak and irritable, and the eyes yellowish.

360. What purpose does the bile serve in the vital economy?

Although the bile is an excrementitious substance, yet at times it serves an important purpose in the vital economy. If there is too much acid in our food, it is sometimes neutralized by the action of the bile upon it, the bile being of an alkaline character. Oily substances which have passed into the duodenum, are acted upon by the alkali of the bile, converting them into a saponaceous substance, which is immediately acted upon by the pancreatic juice, and other chymifying agents, and with difficulty converted into chyme. When there is a considerable amount of fatty matter connected with our food, it cannot be so far chymified in the stomach as to pass with safety into the duodenum. The stomach becomes irritated with its unmanageable contents, and through the organic nerves a sympathy is created in the biliary apparatus, which pours its bile freely into the duodenum, and, instead of the bile taking its usual descending course, it is carried up and admitted through the pyloric orifice to convert the oil and fat into a saponaceous substance, that they can be acted upon by the juices of the stomach. This necessary introduction of bile into the stomach is contrary to the perfect functional integrity of that organ, and it seems to me that all will say, it is better to let these oily foods alone, than to have a chandler's shop in the stomach.

361. What purpose is served by Bruner's, Lieberkun's, and Peyer's glands?

They secrete from the blood, and pour into the duodenum, different kinds of juices, which serve their purpose in the chymifying process. It is supposed that the intestinal juice from Lieberkun's glands turns the starch of the food into sugar.

362. What is the office of the duodenum?

To further the chymifying process commenced in the stomach, and prepare the nutritive portions of food to pass into the lacteals of the intestinal tube as chyle.

363. How is the action of the duodenum accomplished?

As the chyme passes from the gastric cavity into the duodenum, it is instantly perceived by the nerves of organic sensibility, and through them the muscles of the part are excited to action, causing a worm-like motion, by the contraction of the muscles from above downward. By this motion the chyme is slowly carried along the intestinal tube, its course being considerably retarded by the folds of the mucous membrane. While the chyme is passing along the tube, a solvent fluid, nearly resembling the gastric juice, exudes from the vessels of the membrane. As soon as the chyme enters the small intestine, the pancreatic, hepatic, and intestinal juices are poured upon it, and such changes are wrought in it, as gradually adapt all the usable parts to pass into the circulation. At the very entrance of the small intestine, the lacteals, which very numerously abound in this section of the alimentary canal, begin to act on the most perfectly assimilated portions of it, and to elaborate from it their peculiar fluid, called the chyle. As the chyme moves slowly along the living tube, the lacteals in that part of the intestine are acting on the most perfectly assimilated portion of the chyme, and at the same time the less perfectly assimilated portions are preparing for the lacteals of the succeeding part.

364. What are the jejunum and ileum?

They are that portion of the small intestine usually called the mesenteric portion. They are merely extensions of the duodenum, with slight modifications. The jejunum is the upper two-fifths below the duodenum. The ileum is the lower three-fifths. It opens into the large intestine at an obtuse angle.

365. What function is accomplished in the jejunum and ileum?

It is supposed that the completion of the work of chylification, and the preparation of excrementitious matter for its passage from the system, is mainly accomplished in these parts.

366. What is essential to the perfect performance of the work of the small intestine, including both chymification and chylification?

In order that this work may be performed with integrity, the stomach should not be employed at the same time. For this reason food should not be eaten between meals, and our meals should be at least six hours apart, and be eaten with regularity. See the chapter on diet.

367. What is the mesentery, and what is its use?

It is the portion of the serous membrane which secures the small intestine. It forms a gathered or folded curtain, which extends from the backbone to the convolutions and arches of the canal. While it holds every part in its relative position, it admits of a full floating motion of the whole.

The mesentery abounds with lymphatic vessels and glands. In these glands the chyle which has been elaborated from the small intestine, passes, and by their action is more and more assimilated to the blood. It is supposed that the chyle, in passing through these glands, has separated from it a portion of the crude substances that may be connected with it. As a large number, if not all, of these lacteals traverse a portion of the liver before pouring their contents into the thoracic duct, it is probable that they there communicate to the venous capillaries the remaining crudities and unassimilated substances contained in the chyle. The thoracic duct, as we noticed in the chapter on the lymphatics, pours its contents into the venous blood just before it enters the right auricle of the heart; from this it passes to the right ventricle, is forced into the lungs, and on returning to the left auricle of the heart, is fitted to enter the general circulation.

368. What is the large intestine?

It is the last portion of the alimentary tube. It is about five feet in length, and is divided into the cæcum, colon, and rectum. The cæcum is the most dilated portion of the intestinal tube. The colon is divided into transverse, ascending, and descending. It makes a remarkable curve upon itself, called the sigmoid flexure. The large intestine commences within the right hip; it then ascends to the liver, turns across the abdomen to the left side, down which it follows, curving over the inner surface of the hip, and then becomes straight. The rectum is the termination of the large intestine.

369. What is the meso-colon?

It is the membranous curtain which holds the

colon in its position, similar to the mesentery.

370. What is the omentum, or caul?

It is the folds of the membranous curtain from the stomach, the arch of the colon, and the liver.

371. What is essential to a healthy condition of the large intestine?

A regular daily evacuation of the accumulations in the colon, a short time after rising in the morning. A healthy condition of the bowels demands, not only a regular discharge each day, and at a regular time of day, but that each discharge be free, easy, and copious, but not watery, and without pain, straining, or irritation. Constipated and irregular action of the bowels, give rise to most of the diseases that may be named. To purge the bowels with physic only leaves them in a worse condition than before. In case of a constipated and fevered condition of the bowels, freely use tepid injections of pure, soft water, and secure at least one good passage at the regular time in the twenty-four hours. It is best, however, to reduce the heat of the abdomen by external applications to the bowels of cool cloths—not cold, except in severe fever. When severe diarrhea occurs, mild sitz baths and cool injections may be occasionally employed to advantage.

372. What habits, aside from irregularities in diet, tend to produce a diseased condition of the bowels?

Tight clothing or bandages just above the hips, will tend to obstruct the passage of the contents of the colon, both from it upward into the transverse, and from it downward into the descending. By such pernicious habits the colon is constricted, fœcal matter accumulates in it, and life is desstroyed as the result. Thousands thus perish every year. To secure a healthy action of the alimentary tube throughout its whole extent, proper care must be had of the colon. Clothing should not be fastened upon the hips, but should hang suspended from the shoulders, and be worn loosely around the body.

### THE SPLEEN.

873. What is the spleen, and what seems to be its office in the human system?

The spleen is situated in the upper and back part of the abdominal cavity, on the left side, between the diaphragm and the left kidney. It is attached to the diaphragm, the stomach, and the ascending colon, in a loose manner, by folds of the peritoneum. It is extremely spongy, and is formed almost entirely of blood-vessels, lymphatics, and cells, woven together by cellular tissue, and surrounded by a very firm sero-fibrous membrane. It seems to be an appendage of the portal system, and its blood-vessels empty into the portal veins. The spleen seems to serve the purpose of a reservoir, to receive a portion of the blood when its volume is increased by a rapid flow through the liver, and to retain it till it can be acted upon by the liver.

### THE KIDNEYS.

874. What are the kidneys, and what purpose do they serve in the system?

They are two similarly-shaped glands, of a dark brown color, about four and a half inches in length by two in breadth and one in thickness, of the form of kidney beans. They are situated on either side of the spinal column, the right one being a little lower than the left. The right kidney is in contact with the liver, the duodenum and ascending colon; the left is in contact with the spleen, pancreas, stomach, and descending colon. The kidneys are constructed of sinewy fibers, woven very densely together, being filled with an immense number of minute tubes, which empty their contents into a tube on each side of the bladder, which discharges its accumulated contents through the urinary ducts. The blood seems to pass through two distinct systems of capillary vessels in the kidneys, in its course from the arteries to the veins. They take up from the blood the water; they also eliminate from the blood saline and waste matters, sugar, albumen, &c.

375. What is essential to a healthful action of the kidneys?

It is important to attend to the solicitations of nature, and not retain for any length of time the accumulated urine. The kind of water we use greatly affects the action of the kidneys. Hard water, holding lime in solution, is liable to cause maladies most painful to be borne. None but the purest soft water should be used for culinary and drinking purposes.

376. What are the supra-renal capsules?

They seem to occupy some relation to the early action of the kidneys. They are two small, yellowish, flattened bodies, surmounting the kidneys and inclining toward the vertebral column.

## Chapter Nine.

### DIET, OR PROPER FOOD AND DRINK.

377. What is essential in selecting proper food and drink for the human stomach?

It is essential to understand the elements requisite to build up the human body; to ascertain, not only what substances *contain* those elements, but also which ones contain them in a condition to be most readily assimilated to the wants of the vital economy in carrying out the building-up process of the system.

378. Does it follow, because a certain substance contains elements that enter into the formation of the human body, that, therefore, that substance will nourish the body?

It does not. There are substances which may yield, under chemical analysis, some of the elements of the human body, whereas the chemical laboratory within the organic domain would fail to find any place in the vital economy for such substances as food, and would simply expel them from the system as waste matter.

379. How many elements have been found in living bodies?

Nineteen.

380. How many of these nineteen elements are regarded as essential constituents of the human body?

Thirteen. These are carbon, hydrogen, oxygen, nitrogen, phosphorus, sulphur, iron, chlorine, sodium, calcium, potassium, magnesium, and fluorine.

381. What articles are supposed to afford nourishment to every part of the body?

Milk to the nursing child, and wheat and apples to those of more advanced life. Yet neither of these yield to chemical analysis all the elements of the human body.

382. How do you account for this, if it is essential to build up every part of the body, that these elements should exist in the food?

It is supposed that the vital economy of the body has a power of transmuting its substances, and as it combines and commingles them, produces substances which could not previously be detected in the food.

383. What seems to be the most natural food for man?

If we reason from the law of adaptation, man was, in his creation, adapted to a diet of fruits, grains, and vegetables. When the Lord had placed man upon the earth, he said to him, "Behold I have given you every herb bearing seed, which is upon the face of all the earth, and every tree, in the which is the fruit of a tree yielding seed; to you it shall be for meat."* His nature must have been adapted to a diet of this kind. Man's perverted appetites may lead him to clamor, as did the Israelites, for the flesh-pots of Egypt; yet his constitutional nature is best fitted to a vegetable diet.

384. What is one of the principal arguments advanced in favor of flesh as food?

The argument is that flesh contains nitrogen, and that this is needed to build up the body. It is true that flesh-meat contains about fifteen per cent. of nitrogen, while, wheat, rye, oats, barley, corn, rice, peas, and beans, contain only from two to five per cent. of nitrogen; yet these articles are about three times as nutritious as flesh-meat. It

---

*Genesis i, 29.

is therefore apparent that something besides nitrogen is needed to build up the body.

385. Is it necessary to eat the flesh of animals to obtain any of the elements of our bodies?

It is not. Phosphorus, which is a constituent of the bony, muscular and nervous tissue of the body, is found in nearly all vegetable substances, in combination with lime or magnesia. Sulphur, which is found in the hair, bones, saliva, &c., is readily detected in white cabbage, potatoes, peas, and other vegetables. Iron, which may be found in exceedingly small quantities in organized beings, is found in small particles in most vegetables used as food, as cabbage, potatoes, and peas. Chlorine, which is found in the blood, in the gastric juice, and the saliva, is a constituent of nearly all vegetable aliments, making it unnecessary to burden our systems with common salt to furnish chlorine to the body. Calcium, which is found in all the animal solids, in the blood, and in most of the secretions, is a constituent of most vegetables, of the cereals, &c. Magnesium and potassium, found in the blood, teeth, bones, and nerves, are constituents of grains, potatoes, grapes, &c.

386. What are the proximate elements of the body?

Those elements that are readily assimilated to the system, are water, gum, sugar, starch, lignin, jelly, fibrin, albumen, casein, gluten, gelatin, acids, and salts. These are all compounded of two or more chemical elements, and are produced in the growth of nutritive plants of the vegetable kingdom.

387. Is alcohol capable of nourishing the body?

It is not. It is the result of the death and putrefaction of organic vegetable matter. It is antivital in its nature. When taken into the stomach, it highly inflames that organ, and by the strong vital reaction, it is expelled from the gastric cavity into the small intestine, and extends its inflammation throughout the whole length of that canal. It always retards chymification, and renders the process less perfect, and diminishes the functional power of the stomach. As a result, it destroys the vital properties and the vital constitution of the tissues of the body. The same is true of all intoxicating drinks, and true also of such narcotics as tea, coffee, and tobacco, although, if used in moderate quantities, their effects may not be so soon perceived as in the case of intoxicating liquors. Nevertheless they slowly undermine the vital power of the system, in the same ratio as they unduly stimulate it.

388. Can the body be sustained by placing nutritious matter only in the gastric cavity?

The stomach and alimentary canal are constructed with reference to the disposition of food containing nutritious and innutritious matter. Their work is to receive such food, at proper times, in proper quantities, after it has been thoroughly masticated and insalivated in the mouth, and completely to dissolve it, or separate its nutritious from its innutritious matter; and convert the nutritious matter into chyme, and present this to the absorbing mouths of the lacteals, and then to remove the innutritious residuum from the organic domain. If only concentrated nutritious matter were placed in the alimentary organs, we should soon destroy the functional power of the organ, and break down the general function of nutrition, and death would soon ensue.

889. What has been shown by experiment in this matter?

That dogs fed on superfine wheat flour bread, and water, will die in about seven weeks; but if fed on bread made of unbolted wheat meal, and water, they will thrive and do well. Horses fed on grain, or meal and water alone, will die in a short time; but mix with the meal, or grain, cut straw, or even wood shavings, and they will do well. Instances might be cited, where horses were being transported on the sea, and their hay being carried away by a storm, &c. While they were fed on grain alone, many of them drooped and died, but on feeding the balance of them shavings of spars and stave timber with their grain, their appetites returned, and they thrived and did well.

890. Is the same true of persons in this respect that you have said of animals?

Yes; children fed for a considerable time on superfine flour bread, sugar, butter, &c., become weak and sickly, and are often covered with sores, and afflicted with scrofulous diseases. But if, instead of this, the child be fed on good bread, made of unbolted wheat meal, with milk and water, or pure soft water for drink, and be allowed to indulge freely in the use of good fruits in their seasons, and in other respects be properly treated, it will be healthy, robust and sprightly. Again, a single pound of good wheat contains about ten ounces of farina, six drachms of gluten, and two drachms of sugar. A robust laboring man may be healthfully sustained on one pound of good wheat bread per day, with pure water, for any length of time he chooses, without the least inconvenience; but, let him undertake to live on ten ounces of pure farina, six drachms of gluten, and two drachms of sugar per day, with pure soft water, and death will terminate his existence in less than a year. Concentrated food, then, is a source of disorder to the digestive organs, and of disease to the whole system.

391. Does the body need stimulating food in order to thrive?

Every foreign substance from which the body can derive aliment, possesses a stimulating quality proportionate in power to its quantity of nourishment. Some substances are more nourishing and less stimulating, others are more stimulating than nourishing. Some substances stimulate without nourishing at all. These last should never be used as articles of food, but only used, if at all, in particular conditions of the system. If a proper amount of stimulus is in the food, the digestive organs are excited to a healthy action. If the excitement is very intense the organ is debilitated, and a painful sense of prostration is felt. That food should therefore be used which is assimilated, and appropriated by the vital functions, with the least expenditure of vital power.

392. What, aside from spirituous liquors, may be considered the most injurious stimulants?

Tea, coffee, tobacco, mustard, cayenne, black pepper, allspice, cinnamon, cloves, mace, nutmeg, ginger, &c. Of all these the best physiological rule to adopt in regard to their use is—"*The less the better.*"

393. What food is supposed by most persons to be the most nourishing and invigorating?

The flesh of animals.

394. Is meat more nutritious than vegetable food?

The ablest and most accurate chemists of the present age have shown by actual experiment that the various kinds of flesh meat average about twenty-five parts of nutriment out of every hundred parts, while rice, wheat, peas, and beans, afford from eighty to ninety per cent. Potatoes, ranking first among the edible roots, afford about twenty-five per cent. of nutriment, being quite as nutritious as meat. A pound of rice contains more nutritious matter than three pounds of the best butchers' meat; and three pounds of good wheat bread contains more than six pounds of flesh; and three pounds of potatoes as much as the same amount of flesh. Farinaceous seeds contain more nutrition than other kinds of aliment, which is probably the reason they have been called the "staff of life."

395. *If flesh meat is less nourishing, may not the innutritious matter connected with it help the work of digestion?*

No; the nutriment connected with the meat is more stimulating in proportion to the amount of nutriment it affords, than a vegetable diet. All the fluids and substances elaborated from blood made from flesh-meat, are more exciting to the parts on which they severally act, and cause a greater rapidity of vital action and expenditure in the whole system, than is effected by the use of pure and proper vegetable food. The pulse in a robust person, who lives on a vegetable diet, is from ten to thirty beats less per minute than that of one living on the ordinary highly-seasoned meat diet. Meat causes a great expenditure of vital power in its digestion, and hence, leaves the digestive organs much exhausted after the performance of their duties. So that, although meat may pass through the human stomach quicker than some vegetables, and consequently has generally been considered easier of digestion, it is actually the most difficult to digest. It is because a greater draft is made on the vital energies to digest meat than vegetables, that a greater degree of exhaustion is felt in the epigastric region, when the food has passed from the stomach into the intestinal canal, and why persons using flesh-meats suffer more distress from hunger when they pass their usual meal hour, than those who subsist on a pure vegetable aliment.

396. What kind of diet has the preference in proportion to its amount of nutrition?

That which exhausts the vital powers the least. Actual experiment has shown, that, although a pound of unbolted wheat-meal bread contains only about three times as much nutriment as one pound of meat, it will actually sustain a man accustomed to such a diet longer and better than four pounds of meat will sustain a man in similar strength, accustomed to meat diet. Persons subsisting on a well-chosen vegetable diet can endure protracted labor, fatigue, and exposure, much longer without food, than they who subsist mostly or entirely on flesh-meat.

397. But, if this is correct, why do those who leave off flesh-eating and subsist on vegetables feel weak and languid when they make the change?

It is because the flesh meat is more stimulating, and that which we suppose to be strength is only actually the whipping-up of our energies under the spur of stimulants. The system, too, may not be accustomed to this kind of diet, and, as it requires a different kind of gastric juice to digest vegetable than it does to digest animal food, a little time is requisite that the stomach may adapt its secretions to the new diet. For this reason a change of this kind should not be suddenly made. While such changes are being made, the person should be extremely careful not to exhaust his energies by over-laboring, either mentally or physically. In all changes of diet, the new kind should be partaken of sparingly at first, and the amount gradually increased, until finally the new may entirely take the place of the old. The reason many persons make themselves sick in using green peas, corn, beans, &c., is not because these articles are themselves injurious, but because they eat largely of them before the stomach and alimentary tube have become adapted to them.

398. What is the effect of causing the body to flesh up quickly and grow rapidly?

It rapidly expends the resources of the vital constitution, increases the danger of disease, and shortens life. The truth is, that which grows quickly decays quickly, or as the old adage says, "soon ripe, soon rotten." The more stimulating the diet, the more rapidly the changes in the structure of the body take place; hence, as meat is more stimulating and heating than vegetables, the development and growth of the body will be slower with the vegetable-eater than with the meat-eater, and his life longer. Not only is this the case, but muscle that is slowly developed has more firmness of texture; hence, greater strength, and more power of endurance. Therefore, muscle formed of vegetable aliment has the preference. The same effect is produced in the development of the nervous tissue. That formed from pure vegetable aliment is susceptible of higher sensorial power and activity than that organized from blood formed of flesh-meat; so the minds of vegetable-eaters—other habits and health being equal—are more cheerful, stronger, and capable of more protracted efforts than flesh-eaters. Not only have we the case of Daniel and his fellows, in confirmation of this fact, but those of Sir Isaac Newton, and scores of others, as well as the great contrast in the intellectual powers of those nations who subsist on vegetables and those who live on meat, oils, &c.

399. What other great objection have you to the use of flesh-meat as an article of diet?

Its great liability to be itself diseased, and thus imparting directly to the blood and tissues of the body scores of diseases with their attendant evils. The very process of fattening animals tends to cause a collection of adipose matter in their bodies. This is a fruitful source of disease to the animal. It is a matter of no uncommon occurrence for hogs to be killed—in Western phrase—"*to save their lives.*" Few animals are prepared for the city markets whose livers, or some other parts, are not more or less affected with disease. The greatest difficulty in the matter is, persons buying such meat know not in what condition it was killed—whether in a heated, over-driven, angry condition; or whether in disease or health. The safest rule to follow, if persons must eat meat, is to select wild game, and avoid all meats not dressed under their own inspection. For myself, I choose rather to live on the products of the vegetable kingdom.

400. What kinds of food are the most objectionable?

Those containing animal fat, oils, and melted

butter, with several kinds of fish, as eels, sprats, salmon, and herring.

401. Why are butter and animal oils and fats, objectionable as articles of food?

They are too concentrated for human aliment. They are but slightly nutritious, and are filled with impure material, comparing very well with venous blood for impurity. They contain about 80 per cent. of carbon, and hence, furnish additional labor to the lungs to expel this from the system. Butter, when taken into the system, takes the form of an animal oil. These oils float upon the top of the food, and remain in the stomach till all the other portions of food have been chymified and passed from the stomach. Hence, they are liable to generate rancid acids in the stomach, and even when the food has passed from the stomach, these oils are not digested until bile—in the unnatural way mentioned in the last chapter—passes from the duodenum to the stomach, and turns the oil into a saponaceous substance before it can be acted upon by the gastric juice, and digested. Butter, if used at all, should be made from the milk of healthy cows, and used only when it is new and sweet, and but very slightly salted. It should never be used in a melted form, nor upon anything hot enough to melt it.

402. If butter is unhealthy, should milk or cream be used?

Milk, although about seven-eighths water, affords more nutriment to the system than butter or hog's lard. Although it is an animal secretion, it cannot really be called an animal food; yet it is more easily affected by the general health and food of the animal producing it than any other secretion of the body. The child may be affected by injurious drugs, stimulants, and improper food of the mother, while she herself experiences but little difficulty. When milk is pure, and from healthy cows, it is the best form of food aside from vegetables, especially for children. And except in diseased states of the digestive organs, its moderate employment is attended with no inconvenience; yet in many diseases it is indispensably necessary to prohibit entirely the use of milk, especially if the disease is of an inflammatory character, or one which increased excitement will aggravate.

403. What is the best milk?

That procured from healthy cows, which, during the season of grazing, run at large in the open field and crop their food from a pure soil; and during the winter are fed on good hay, and, if housed at all, are kept in clean and well-ventilated stables, and every day thoroughly curried and cleaned, and supplied with pure water for drink, and suffered to take regular exercise in the open air.

404. What do you say of cream?

When it is sweet it is perfectly soluble in water, and mixes freely with the fluids of the mouth and stomach; and therefore, if it is free from any deleterious properties, is far less objectionable than butter as an article of diet.

405. What of cheese as an article of diet?

Cheese is always more or less difficult of digestion, beside being frequently colored by poisonous substances, as annato, arsenic, &c. Old cheese should never be used. Cheese not more than three months old made of milk from which the cream has been mostly taken, is most easily digested. But, of cheese in general, it would be well for all to keep in mind the old adage,

> "Cheese is a mighty elf,
> Digesting all things but itself."

Old cheese is exceedingly obnoxious as an aliment.

406. What of curds, and Dutch cheese?

Curds made of fresh milk, and pot-cheese made of milk as soon as it sours, before it becomes bitter, are not very objectionable.

407. What of soups?

Flesh soups are very objectionable forms of animal food. Soups in general are too complicated to be healthy. Flesh broths are simply water holding in solution the nutrient particles of the flesh meat in a very concentrated form. The first process of digestion, when this food is received into the stomach, is for the absorbents to take up all the water it contains; and as it has been swallowed quickly without mastication, and unmingled with saliva, it is left dryer than aliment which requires chewing, and hence is difficult of digestion.

408. What have you to say of fish?

The flesh of healthy fish is less exciting, and less nourishing than beef, mutton, and other animals. Smoked fish, and in fact, salt smoked meats of any kind are very difficult of digestion. Salt fish should not be eaten. Fresh scale fish, recently taken from the ocean, or from rivers of pure water, or from running streams, or from lakes with inlets and outlets of running water, are the least objectionable of any. All shell-fish, including oysters, are objectionable as articles of diet. They contain only about 12½ per cent. of solid matter, but very little nutrition, and are very difficult of digestion. They digest the most readily when eaten raw. Lobsters, and all fish not having fins and scales, are hard of digestion. We suppose they were forbidden in the instruction given to Moses because they were not adapted to man's nature.

409. Are eggs a proper article of food?

Eggs are more animalized, and more exciting to the system than milk. Eggs, fresh and good, raw or rare-boiled, without the use of fat or oily matter, are moderately nutritious, and easy of digestion. Poached eggs are very pernicious. Eggs hard-boiled or fried, are extremely difficult of digestion. Persons whose diseases are such that they cannot use milk, may suffer the same inconvenience from the use of eggs, and would do better to let them alone.

410. Is the eating of common salt any great advantage to the human system?

It is not, being innutritious, and indigestible. It is irritating, and its presence in the system tends to produce chronic debility, and disease of the stomach, intestines, absorbents, veins, heart, arteries, and all other organs of the system, retarding those functions by which the vital changes are effected. If, then, salt is used at all, it should be used very sparingly. It is argued that deer and some other herbivorous animals go in search of salt and brackish water. It should be observed by those raising this objection, that these animals who frequent salt water pools only do so in warm weather, and even then only seek it when they are diseased by worms, bots, or grubs in the alimentary cavity. So it is not sought by them as seasoning to their food, but merely as a medicine.

411. What are considered to be the organic acids?

The acetic, citric, tartaric, malic, oxalic, and lactic. They all exist in those fruits and vegetables which nature has provided for our nourishment, except the acetic acid, which, like alcohol, seems to be the result of decay, instead of formation. It is a matter of doubt with many authors whether acetic acid is really an organic acid in the human body.

412. If acid is needed in the system, is not vinegar healthful?

Vinegar, like alcohol, is a product of fermentation, and it is very debilitating to the human stomach.

413. What fruits are considered most wholesome?

Apples, pears, quinces, peaches, nectarines, apricots, cherries, currants, gooseberries, whortleberries, cranberries, grapes, strawberries, raspberries, blackberries, oranges, lemons, limes, citrons, melons, squashes, pumpkins, figs, tomatoes, mulberries, pine-apples, &c.

414. What vegetables afford the purest aliment?

The seeds are wheat, oats, barley, rye, rice, Indian-corn, peas, beans, and various kinds of nuts. Among the roots are the potato, turnip, carrot, beet, parsnip, artichoke, &c. Onions, leeks, asparagus, cabbage, and many other leaves, are considered wholesome.

415. What is the best condition in which fruit can be eaten?

As nearly as possible, whether raw or cooked, it should be eaten in its natural state, free from spices. It should also be eaten at our meals, with our food, as a portion of the meal, and not between meals, or as a dessert after we have eaten sufficiently of something else.

416. Aside from the healthful building up of the system, is there any peculiar benefit experienced by the vegetable-eater?

Careful experiment has shown that the bodies of those that subsist on a vegetable diet, their other habits being right, are not only healthy, but their bodies are but little affected by prevailing epidemics, and contagious diseases. So we may conclude that the systems of such are better prepared to resist disease than that of the meat-eater.

417. Is sugar healthful?

In the shape of candies it is very unwholesome. Sugar is highly carbonaceous, affording but little nourishment to the system. Most of the refined sugars have a constipating effect on the bowels. The best article of sugar is a pale yellow, with large, clear, brilliant crystals. A good syrup can be made by dissolving two pounds and a half of sugar in a pint of pure soft water.

418. What have you to say in regard to the proper amount of food to be eaten?

Care should be taken to eat no more than is needful to meet the wants of our bodies. To place in our stomachs a great amount of food imparts no more nutrition to the body, while it imposes upon it an extra amount of labor, debilitates the stomach and produces indigestion. An active, out-of-door laborer can dispose of more food than those of sedentary habits. By carefully watching their own feelings all can readily decide on a proper amount of food to sustain their bodies without producing a sense of oppression. Persons who have passed the time of their meal and get very hungry are liable to indulge in excessive eating. In such cases do not allow yourself to eat any more than at an ordinary meal.

419. What should be the general habit at table?

Never take the meal in a hurry. Chew the food thoroughly, and swallow it slowly. Let the table be a place of pleasantry, cheerfulness, and social enjoyment. Ever remember the saying of Lord Bacon; "If you would live long and enjoy life, be cheerful at your meals, and on going to bed."

420. What are the proper times of eating?

It was said in the last chapter that food should not be introduced into the stomach oftener than once in six hours, not meaning to be understood, however, that the human stomach needed food every six hours in the twenty-four, but that, whether we take two or three meals, they should be at least six hours apart. We should scrupulously, and rigidly avoid partaking of the least particle of aliment at any time except with our regular meals. Those of strong digestive organs, may not, for a time, discover any inconvenience in eating between meals, yet this is set down by the best of authors as one of the most fruitful causes of dyspepsia. Sooner or later those who eat thus at random, will have all the horrors of dyspepsia upon them. To avoid this, they must turn from this evil practice.

421. How many meals a day should we eat?

Our meals should be taken, as nearly as possible at the same hours each day. They should not be so far apart as to cause us to overeat. They should not be so near together as to introduce food into the gastric cavity before that already there is digested. The welfare of the human constitution requires at least two regular meals in the twenty-four hours. Those accustomed to a vegetable and fruit diet, who are of sedentary habits, can very soon accustom themselves to two meals a day without any inconvenience. These meals should be at about 7 A. M., and at 1 or 2 P. M. Farmers who have become accustomed to the change will experience no inconvenience in restricting themselves to two meals a day, if they are temperate in the number of hours they labor. If farmers think they must have three meals, the best times to take them, are at 6 A. M., 12 M., and 6 P. M. In this case their supper should consist of a small amount of plain simple food. Those however who find their sleep disturbed, and their stomachs and mouths tasting bad in the morning, would do better to resort to moderation in their labor, and abstain from their third meal.

422. How often should nursing children be fed?

Once in three hours is as often as children should be nourished. When they are old enough to eat solid food, they should eat three or at most four regular meals a day until they are two or three years of age, when they may safely be brought to two meals a day. They should never be allowed to take a morsel of food except at their regular meals.

423. What is the best substitute for the mother's milk for a nursing child?

Nothing should be substituted, if the mother be healthy; but if substitutes must be had, take good wheat, clean and dry, but let it be ground without bolting, put a tablespoonful of this into a pint of pure water, boil it about fifteen minutes, and add to it one pint of good, new milk from a young healthy cow, fed on hay or grass. The milk should not be changed for that of any other cow, if it can be avoided. When children are weaned, good coarse wheat bread, good new milk diluted with about half as much pure soft water, boiled, and a proper supply of ripe fruits in their seasons, should constitute their diet.

424. How should we proceed in case of loss of appetite?

If we have no desire for food at the regular time for a meal, it is better to nearly abstain from food till the next regular meal-time arrives.

425. After fasting, how shall we proceed?

Eat lightly of food at the first meal and return gradually to the accustomed amount of food. The moderation of the return to the usual amount may be in proportion to the length and severity of the fast.

426. What is the most wholesome drink for man?

Filtered soft water. This should be used, not to wash down the food, but to quench the thirst. Water should be *cool* but not very cold, or iced. Hot water is very debilitating and weakening to the stomach. Beer, soda, and all kinds of mineral waters should be carefully avoided. Excessive drinking, even of pure water, is weakening to the system, as it imposes additional labors upon the absorbing and eliminating organs. Food of an exciting and irritating kind, as that filled with salt and other seasonings, will call for excessive drinking, and for these reasons all such food should be avoided.

427. What is the best habit to observe in regard to drinking water?

The best time for drinking water is undoubtedly when the stomach is empty—on first rising in the morning, and half an hour or an hour before meals, or three or four hours after a meal. There may be conditions of the stomach which require drink at meals, but, as before said, this should be done when the food is out of the mouth, and not to wash down the food. Those who have weakened their digestive powers so that their stomachs will not bear cool water, should commence with a small quantity, gradually increasing the amount, and thus accustom themselves to its use.

## Chapter Ten.

### MISCELLANEOUS ITEMS.

428. What has been the common theory in relation to the production of animal heat?

That the production of animal heat was a mere chemical process; the lungs serving as a stove, or fire-place, and the carbonaceous substances of the food serving as fuel, "to be burned in the lungs." If this were the correct theory, corpulent persons, who are surcharged with carbon, should bear cold better than lean persons, who have but little of it in their bodies; but such is not the fact. Again, the carbonic-acid gas, which is expelled from the lungs, instead of being produced *in* the lungs is really—a large portion at least—formed in the tissues of the body distant from the lungs.

429. What would be the fact, if the above theory of animal heat were correct?

That those eating *fat, blubber, oil,* &c., would have the greatest amount of heat in their bodies, and hence be the best prepared to resist cold. But the facts in the case are just the reverse: the man who is fully accustomed to a pure vegetable diet, can endure severe cold, or bear the same degree of cold much longer, than the man who is fully accustomed to a flesh diet. In the coldest parts of Russia, the people subsist on coarse vegetable food, and are hardy and vigorous. The same is true of Siberia. Exiles there, accustomed to a vegetable diet, endure the severities of the climate the best.

430. How is animal heat produced?

It is probably a vital function, depending immediately on the vital properties and functional powers of the nerves of organic life. Heat is probably not peculiar to any particular part of the system, but is as universal as the distribution of the nerves of organic life and the blood-vessels. The combination of oxygen with carbon in all the tissues of the body is undoubtedly a source of animal heat, in common with all the organic functions and chemical changes which take place in the body. The great regulator to the heat of the body is a healthy skin.

431. What do you say in general of the use of fire in heating the body?

Its effect is to relax and debilitate the system, and diminish the power of the body to regulate its own temperature. As far as possible we should let our bodies be warmed by their own vigorous calorific function. Let fire be used only as a necessary evil. Do not expose the body to unequal temperatures at the same time, or to powerful heat on one side, and cold on the other. Let the temperature of your room be mild and equal in every part, ranging fram 55° to 70°, according to the health and bodily vigor of the person.

432. What is requisite to the due regulation of the animal temperature?

Good digestion, free respiration, vigorous circulation, proper assimilation, and perfect depuration, or in short,—*good health.*

433. What further is essential to the proper performance of the organic functions, and the generation of animal heat?

That the body should be regularly and systematically exercised. The health of the whole body depends on the fluids of the body being kept in constant motion; that the grand vital circulation may be kept up, a proper supply of blood be carried to every part, and the vital changes be accomplished with perfectness. If exercise is neglected, the body will become feeble, and all its physiological powers will be diminished; but if exercise is regularly and properly attended to, the whole system will be invigorated, and fitted for usefulness and enjoyment.

434. What do you say in relation to proper times of exercise?

The most severe and active exercise should never be performed on a full stomach, nor immediately before or after a meal. A laboring man who takes his meals at the regular hours we have recommended, and retires and rises at regular and early hours, would do well to exercise moderately an hour or so before breakfast, perform their hardest labor between breakfast and dinner, and work moderately again after dinner. Much evening work is a violation of nature. Persons of sedentary occupations should choose such out-of-door exercises as they can habitually and regularly attend to. Their most active exercise should take place at those times in the day when the stomach is partially empty. Evening exercise is not objectionable for them.

435. What is essential to secure the full benefit of exercise?

It is essential that it should be coupled with an object of either utility or amusement. Some useful business pursuit which requires, and hence secures, attention and labor during several hours of the day—according to the strength—is essential to the best sanitary condition of the body. If your exercise is simply walking, don't go into it with a monotonous drudge, but have some object before you, and let your mind drop the train of thought which has sufficiently taxed it. Walk with some one; walk with cheerfulness, and intersperse it, if strong enough, with running a short race, seeing who can run or pace to a given point the quickest. Four hours of daily labor on the soil is probably the best thing to invigorate the body and mind of those of sedentary habits.

> "Other creatures all day long
> Rove idle, and unemployed, and less need rest:
> Man hath his daily work of body or mind
> Appointed, which declares his dignity,
> And the regard of Heaven on all his ways;
> While other animals inactive range,
> And of their doings God takes no account."

436. What is the comparative difference between those who exercise, and those who do not?

If man takes too little voluntary exercise, he suffers; if his exercise is too excessive, he also suffers. The sufferings of excessive exercise bear no proportion to those resulting from inactivity. A man greatly abbreviates his life by over-toiling, and yet, for the most of his life, may have a cheerful mind, good health, and sweet sleep. But a lack of exercise, connected, as it often is, with excessive alimentation, and other dietetic errors, produces the most intolerable of human miseries. Temperance in this matter, then, will promote health, happiness and length of days.

437. What is especially essential to the health of children?

The welfare of the body, especially of children, demands that each part should be duly exercised. Children are instinctively inclined to action, and the symmetrical development of their bodies requires much exercise in the open air. It is unnatural and improper to keep them in a state of confinement or inaction for any considerable time. Girls should be allowed to exercise as freely, in open air, while their bodies are growing, as boys. Avoid rocking young children to keep them quiet, or get them to sleep. When they begin to notice things, give them plenty of room on the smooth floor, with plenty of playthings, anything with which they will not injure themselves.

438. What of the exercise of aged people?

Aged people, after they have retired from the active employments of life, must keep up their regular exercise, or they will soon become feeble and infirm. Walking and horse-back riding are among the best modes of exercise for the aged.

They also should connect cheerfulness with their exercise. Exercising the mental, as well as the bodily powers, serves in no small measure to preserve life and enjoyment with the aged.

### DISEASE.

439. What may disease in general terms be called?

Disease in its incipient state, as a general fact, may be said to be nothing more than an excessive healthy action of nature in resisting morbific or irritating causes. This action continued too long brings the overacting part into a morbid condition, and may involve the whole system in sympathetic irritation. All that nature asks to cure such conditions is to have the disturbing causes which produce this morbid condition removed. Chronic diseases result, if the disturbing cause is not removed, in the change of structure of the parts of the system acted upon. If the disturbing causes are removed before change of structure takes place, the diseased action of the part will not long continue, but nature will at once commence its restorative work.

440. What are the most fruitful sources of disease?

In general, diseases are produced by bad air, improper light, impure and improper food and drink, excessive or defective alimentation, indolence or over-exertion, and unregulated passions.

441. What, then, are the conditions of the body in disease?

Impure blood, unhealthy secretions, obstructions in the capillary vessels, excessive action in some parts or organs, with deficient action in others, unequal temperature, or in other words, a *loss* of balance in the circulation and action of the various parts of the vital machinery, producing great discord in some portions of it, and more or less disorder in all.

442. Is pain always manifest in the part affected?

It is not. It requires considerable skill and experience to tell, in all cases, by the pain, just where the disease is located. Pain produced by some diseases of the liver is indicated by a pain under the shoulder. Hip disease is frequently first perceived by a pain in the knee. A morbid condition of the stomach and liver may be perceived first by pain or soreness in the head.

443. What is the first effort to be made in paving the way for the restoration of natural action in the body?

To remove obstructions, wash away impurities, supply healthful nutriment, regulate the temperature, relax intense, and intensify torpid action. Water with its proper accompaniments, air, light, food, temperature, exercise and rest, must best answer nature's demands in this work.

444. Can medicines of any kind effect what the system needs in case of diseased action?

Medicinal drugs can never accomplish this work. They may suppress a symptom, remove a pain, transfer an irritation, excite a new vital resistance, produce another obstruction, and so divide the organic struggle between two points, and diminish vital power to resist disease and cause pain, by increasing vital expenditure. Let it always be borne in mind that all medicine is in itself an evil, and its own dire effects on the living body are in all cases, without exception, unfriendly to life. In every case, to a greater or less extent, it wears out life, impairs the constitution, and abbreviates the period of human existence. Still, in the present condition of human nature, there are some cases of disease in which medicine, to some extent, is indispensably necessary to the salvation of human life, but in these cases it must be regarded as a necessary evil. That physician who uses the least medicine is the best friend to man's physical interests.

445. *What can you say of the effect of the passions on the general health of the body?*

All excessive mental irritations, as anger, grief, and despondency, are injurious to the body. In about the same ratio do lustful feelings and habits debase man and undermine his health. The young man or woman who would have a symmetrically-formed body, strength of mind, and a constitution capable of enduring the labors incident to even a temperate life of man or womanhood, should avoid the secret evil habits and vices of youth with the same dread that they would the bites of the most venomous reptiles. Do not sin against God and your own soul by destroying yourself.

# INDEX.

N. B. *The numbers used in this index refer to the number of the question, and not to the page, unless so specified with the number.*

ANATOMY, use of the term, 5; how treated in this work, 6.
Adam's apple, 314.
Atlas, what, and why so called, 51, 52.
Appetite, how to treat the loss of, 424, 425.
Arteries, description of, 168–171; muscular coat of, and its use, 171; number of, 172; where placed, 173; how to know when one is cut, 174; how to check blood from, 174; course of in the system, 175; illustration of, 177, *Fig. X*, p. 71; more branches of than main tubes, 178; connection between them, and why, 179; pulmonary, and their use, 180, 181; termination of, 182.
Air, impure, 327.
Acids, organic, 411; vinegar, 412.
Aliment, canal of, 334; the mouth, as a part of, 335; description of alimentary tube, 351; how this tube is lubricated and preserved from injury, 352.
Animal structures, formation of, 22.
Auricles of heart, 154; necessity of two sets, 160.
Alcohol, how produced, and how it affects the body, 387.
Arbor vitæ, 261.
Absorbent vessels, 219, 223; action of, 224.
Abdominal cavity, its contents, 24, 331, 332; *Fig. XIX*, p. 159; viscera of, 330.

BATHING, benefit of, and rules for, 305, 306.
Brain, where situated, 247; size and construction of, 248, 249; compared with size in other animals, 249; two in man, and reason why, 250; illustrated, *Fig. XIV*, p. 105; explanation of Figure, 251; not all in the head, 252; coverings of, 255; parts of, 256, 257; function of, 265; effect when injured, 266; inactive in sleep, 267; strengthened by proper use, 271; how protected from jars, 272; surface of brain, 273.
Breathing, process of described, 320.
Bile, its use in the system, 360.

# INDEX. 207

Bones, how formed, 32, 34; marrow of, 33; strength of, 36; how weakened, 37; how injured, 38, 44; position and number of, 40, 41, 47; kinds of, 42; in old persons, 45; first one hardened, 43; illustrated, 45, *Fig. I*, p. 24; weight of, 46; of face, 60; of ear, and use of, 61; of tongue, 62; of upper extremities, 95; of lower extremities, 98; sesamoid, 99; in pairs, 100; why mostly cylindrical, 111; dislocations of, 113.

Body, how composed, 25; cubical size of, 26; water principal element of, 26; how often undergoes a change, 39.

Blood, what formed from it, 16; composition and color of, 200–202; amount of in the body, 203; quality dependent upon the food, 208; how carried through the veins, 197; amount passing through the heart per hour, 166; purification of, 325; illustrated, *Fig. XVIII*, p. 153.

Bowels, habits injuring action of, 372; when should be evacuated, 371.

CARPUS, 96; illustrated, *Fig. III*, p. 37.
Cartilages, where placed, 109.
Capillaries, where placed, 183, 187; size and number of, 184; function accomplished in, 185; how blood is circulated in, 186.
Caul, 370.
Carbonic-acid gas, destroys life, feeds vegetation, 21.
Chest, 88.
Chemical elements of body, 29.
Circulation, organs of, 151; central point of, 152; when, how, and by whom, discovered, 194; how to help, 197; time of, in the system, 204, 205; health is a healthy circulation, 206; how to secure good, 207.
Cold, who endure best, 429.
Colds, cause and cure of common, 307, 308.
Clothing, rules for, 309, 372.
Chyle, what produced from, 15; how produces various structures, 22; what it is, 221.
Chyme, how produced in the stomach, 346, 347; what essential to good, 366.

DIAPHRAGM, action of, 322; illustrated, *Fig. XVII*, p. 151.
Diet, effects of change, and why weakening, 397: care necessary in change of, 397.
Digestion, time required for food in the stomach, 348; of air in lungs, 323; how accomplished, 323.
Duodenum, 353; appendages of, 354; office and action of, 368.

# INDEX.

Drinking, with food, 339, 427; wholesome and unwholesome, 426; best habits in regard to, 427.

Disease, what it is, 439; the most fruitful source of, 441; condition of body when in, 441; how to cure, 443; effect of medicine in, 444.

EAR, construction of, 291; action of, 292; use of, 293; diseases of, 294.

Enamel of teeth, use of, 69; evil of breaking it, 70.

Excretions of the body, 28.

Exercise, muscular, 144; healthful kinds of, 145; position of body in, 146; time of, 282, 434; necessary, 433; how to take, 435; inactivity compared with, 436; for children, 437; for aged people, 438.

Eye, described, 287; how moistened, 288; tears of, how caused, 288; illustration of action of, 289; care of, 290.

Elements, of food, 377, 378; number of, found in human bodies, 373; number that are constituents of the body, 380, 381; proximate of the body, 386.

Employment essential to health, 283.

FASCIE, position of, 148; two kinds of, 149, 150.

Fat, no excess of in health, 24.

Fasting, how proceed after, 425.

Faculties of living bodies, compared with those of vegetables, 14.

Flesh, not necessary as food, 384, 385; not the most nourishing food, 393, 394; cause and effect of increase of in the human body, 398.

Felons, where originate, and how to serve, 113.

Fever-sores, where seated, 113.

Fluids of the body, 28.

Fruits, kinds wholesome, 413; best condition to be eaten, 415.

Food, good, essential to good blood, 208; how acted upon by stomach, 344; change in, effected in stomach, 346; what proper, 377; natural for man, 383; arguments of flesh-eaters on, 384, 385; must not be all nutrition or too concentrated, 387–390; stimulating not necessary, 391; flesh-meat not the most nourishing, but stimulating, 394, 395; vegetable, its benefits, 396; diseased, 399; most objectionable kinds of, 400; why butter, animal oils and fats are objectionable as food, 401; what kind of butter used, if any, 408; of milk or cream as food, 402–404; cheese and Dutch cheese as food, 405, 406; of soups, 407; of fish, 408; of eggs, 409; of salt, 410; of organic acids, 411; proper amount of, 418; how to eat, 419; time of eating, 420, 421; for nursing children, 422, 423.

GANGLIA of the brain, 262.
Glands, salivary, their position, office and action, 335, 336; Bruner's, Lieberkun's and Peyer's, and their office, 354, 361.
Gastric juice, properties of, 338.
Gall bladder, where found, 357.

HAIR, description and care of, 302.
Happiness, true, how secured, 8–10, 13, 276.
Head, number of bones in, 58.
Heart, the central point of circulation, 152; described, 153; a double organ, 154; auricles of, 155; ventricles of, 156; valves of, 157, 158; course of blood through, 159; contractions of, how caused, 161; acts continually, 162; how it rests, 163; number of beats per minute, 164; power of the heart, 165; amount of blood passing through it per hour, 166; illustration, *Fig. IX*, p. 67, explained, 167; diseases of, 209.
Heat of body, how produced, 428–430; who endure cold the best, 429; fire heat, how to be used, 431.
Hygiene, meaning of term, 4.
Hunger, when manifested in healthy state of the stomach, 347, 427.

INSANITY, causes of, 280.
Intestine, small, 351; large, and course of, 368; what essential to health of, 371.

JEJUNUM and ileum, description of, 364; their action, 365.
Joints, position of, 101; number in the body, 102; motions of, 103; construction of, 104; how held together, 105, 106; kinds of, 110.

KIDNEYS, description and use of, 374, 375; capsules of, 376.

LACTEALS, description and action of, 215, 216.
Larynx, its construction and action, 313, 314.
Life a forced state, a battle against causes of death, 19.
Ligaments, 105; position of, 106, 107; consistency of in different periods of life, 108.
Liver, description of, 357, 358; office of, 359; indications of inactive, 359.
Lungs, structure of, 318, 321; mucous membrane of, 318; arteries and veins of, 318; capacity of, 321; care of, 326; exercise of, for students and in-door laborers, 326; diseases of, 329.
Lymphatics, description of, 210; office of, 211; origin and course of, 212; glands of, 213; construction of, 214; num-

ber and kinds of, 217; center of the system of, 218; how related to the venous system, 220.

MASTICATION, evils of improper enumerated, 337.
Mesentery, description of, 215, note, and 367; office of, 333.
Mesocolon, description and use of, 369.
Medulla oblongata described, 260.
Medicines, effect of on the system, 444.
Mind affected by body, and how, 10, 277; what it is, 274; despondency of, 277, 280; comparison of, in vegetable and flesh-eaters, 398.
Morals, correct depend on health, 10.
Mouth, office of in alimentation, 335.
Mucous membrane, in lungs, 318.
Muscles, only element of motion in the body, 23; how act, 144; how composed, and shape of, 115–117, 122; number of, 118; how much of the body is, 119; how moved, 120, 121; appearance of fibers of, 123; number of kinds of, 124; never restored when destroyed, 125; muscles of head and face, 126; illustrated, *Fig. V.*, p. 47; different groups of in the body, and their use, 127–132; muscles at terminus of bowels, 133; of leg, foot, arm and fingers, 134, 135; contraction of, 136; illustrated, *Fig. VII.*, p. 53; vessels nourishing, 138; pairs of, or antagonists, 139; decomposition of, how supplied, 139; of alimentary canal, 140; how alimentary act, 140; how to strengthen, 140; disadvantageous action of, 141; illustrated, *Fig. VIII.*, p. 56; force used in action of, 141; rapidity of movement, 142; how to keep healthy, 143; exercise of, 144; use of the system of, 147.

NAILS, growth, use, and care of, 308.
Nerves, system of, what it is, 225; the man, 225; vitality of, 226; in all organized bodies, 227; of trees, plants, &c., 227; number of systems of, 228; structure of, 229; size of component parts of, 230; center of energy of, in the body, 231, 232; solar plexus of, 232; branches of solar plexus, 233; organic nerves with their plexuses, or presiding centers, 235, 237; effect of suspending the action of organic and animal, 238; orders of organic, and their use, 239; distribution of, 240; nourishment of, 241; when the mind is conscious of the action of organic, 242; ganglionic system of, 243; illustrated, *Fig. XIII.*, p. 100; sympathy of stomach with organic, 244, 245; cerebro-spinal, 246; largest in the body, 253; use of, 254; order of cerebro-spinal, 258; cranial, 260–263; illustrated, *Fig. XV.*, p. 109; description of nine pairs of cranial, 263; description

of thirty-one pairs of spinal, 264; connecting link of the two systems of, 278; derangement of organic system of, first affects the stomach, 279; consistency of, 281.

NITROGEN, proportion in air, and in what proportion supplied to the system in the lungs, 323.
Nourished, how plants and bodies are, 14.
Nutrition, nature requires innutritious food with, and why, 388–390.

ŒSOPHAGUS, description and use of, 341.

ORGANS, thoracic, 311; of voice, 312.
Omentum, office and position of, 333, 370.
Ourselves, how benefited by the study of, 8.
Oxygen, principle that nourishes life, 21.

PAIN, not always where the disease is seated, 442.
Passions, effect of on health of the body, 445.
Pancreas, structure and office of, 355, 356.
Periosteum, description of, 112; its use, 112; seat of felons and fever-sores, 112.
Pelvis, position and use of, 97.
Pleura, what is the, 24, 319.
Position, evils of wrong, in sitting, walking, &c., 146.
Phrenological divisions of brain, 373; organs not alone an index to character, 275.
Physiology, analysis of the word, 1; of what it treats, 2; animal, vegetable and human, 3.
Pyloric orifice of stomach, and its use, 345.

RADICLES, description, position and action of, 191.
Rattle-snake, blood harmless, same blood secretes poison, 15.
Relation of man by his faculties, 7; how that should be maintained, 12, 13.
Recreation, necessity of periods of, 283.
Ribs, form of, 90–92; use of, 93; care of, 94.

SALIVA, glands of, and their position, 335; office and action of, 336.
Swallowing, how effected, 342.
Security of parts of body, 24.
Sternum, how formed, 89.
Sesamoid bones, 99.
Spleen, construction and use of, 373.
Senses, external organs of, 284; of smell, 285; what essential to healthy action of smelling, 286; of sight, 287; of hearing, 291, 292; of taste, 295, 297; of touch, 298–301.

## INDEX.

Sleeping, apartments, 270; who require least, 268; cases of Wesley and Webster, 268; proper amount of, 267; best time for, 269; in morning, and after meals, effect of, 269.
Sweating, excessive, injurious, 304.
Spine, how injured, 55; marrow of, 253; branches of marrow of, 254; construction of marrow of, 259.
Skin, structure of, 299; appendages of, 302; action of the internal, 302; office of internal, 304; care of, 305.
Sigmoid flexure, description of, 351, 368.
Stimulants, admissible, and inadmissible, 391, 392.
Solids of the body, 28, 31.
Stomach, description of, 343; how acts upon food, 344; why, and how kept healthy, 349, 350; how affected by organic nerve derangement, 279; disease, or disturbance of, affects the mind, 280.
Substances of the body, 29, 30.
Skull, parts of, 56; cavities of, for the eyes, 57; bones of, and how secured, 59.
Study, best time for, &c., 282.
Sugar, effect of on system, most wholesome kind of, 417.
System, human, described, 11.
Sympathy, strong between stomach and other organs, 349, 350.

TASTE, organ of 295; how affected, 296; healthy, 297; perverted, 297.
Transmutation of substances in the body, 382.
Trachea, description of, 317; how acts when food is swallowed, 317.
Tendon, Achilles, why so named, 137.
Teeth, of what composed, 29, 62; illustrated, *Fig. II*, p. 29; two sets, 63; reason of two sets, 74; difference in the two sets, 75; number of, 64, 72; how formed, 65–68; vessels and nerves of, 71; care of in a child, 73; different kinds of, 76; use of, 77–79, 81; preservation of, 80; whose teeth best, and why, 82, 87; how care for, 83, 85; toothache, cause of, 84; cause of decay of, 86; when decay fastest, 86.
Temperature, of blood always the same, 20; of rooms, what degree, 431; evils of uneven, 431; aids in regulating animal, 432; exercise essential to proper, in the body, 433.
Tonsils, description and use of, 340.
Thorax, its contents, 24; thoracic duct, 218.
Touch, sense of, where located, 298; where most acute, 301; how acts, 301.

VERTEBRÆ, their use, 50; column of, 48, 49; how constructed, 53, 54.
Veins, structure and position of, 189–191; course of, 192;

difference between arteries and, 193; valves of, 194; three classes of, 195; deep and superficial, 196; pulmonary, 198; of the portal system, 199.

Vegetables, what kinds best, 414; benefits derived by those living on, 416.

Ventilation, necessity of, 327; what essential to good, 328.

Ventricles, two sets of, 154; necessity of two sets, 160; of the brain, 262.

Vital economy, powers of, 15, 16; vital force, or constitution, 17; cannot be increased, 18; may be wasted, 18.

Viscera, three cavities of, 310; illustrated, *Fig. XVI*, p. 143; abdominal, 330.

Vices, secret, destructive to the body, 445.

Vinegar, unwholesome, why, 412.

Vomer, 110, and note.

Vocal cords, 12; ligaments, 314.

Voice, how produced, 315; training of the, 316.

WATER, how absorbed in the stomach, 219.

# GLOSSARY.

AN EXPLANATION OF MEDICAL TERMS USED IN THIS WORK, AND IN MEDICAL WORKS GENERALLY.

*Abdomen.* The lower belly, or that part of the body which lies between the thorax and the bottom of the pelvis.

*Ablution.* Cleansing by water; washing of the body externally.

*Abortion.* A miscarriage, or producing a child before the natural time of birth.

*Abscess.* A cavity containing pus, or a collection of matter; as a comman boil or felon, or any swelling that has come to a head.

*Absorbent.* In anatomy, a vessel which imbibes; in medicine, any substance which absorbs or takes up the fluids of the stomach and bowels.

*Accoucher.* A man who assists women in child-birth.

*Accuminate.* Taper-pointed; the point usually inclines to one side.

*Acetabulum.* The socket that receives the head of the os femoris, or thigh bone.

*Acid.* Sour; sharp or biting to the taste, as acetous acid, or vinegar; citric acid, obtained from lemon, etc.

*Acidity.* The quality of being sour; tartness, or having a sharpness to the taste.

*Acrid.* Sharp, pungent, bitter; biting to the taste.

*Actual Cautery.* A surgical operation, performed by burning or searing with a hot iron.

*Acupuncture.* A surgical operation, performed by pricking the part affected with a needle.

*Acute.* Sharp, ending in a sharp point; acute diseases are of short duration, attended with violent symptoms; it is opposite to chronic.

*Adhesive.* Sticky, tenacious, apt or tending to adhere.

*Adhesive Plaster.* Sticking plaster.

*Adhesive Inflammation.* That kind of inflammation which causes adhesion.

## GLOSSARY.

*Adjuvant.* An assistant; a substance added to a prescription to aid the operation of the principal ingredient or basis.

*Adult Age.* A person grown to full size or strength; manhood or womanhood.

*Affection.* Disorder, disease, malady.

*Affusion.* The act of pouring upon or sprinkling with a liquid substance.

*Albumen.* The white of an egg. A principle of both animal and vegetable matter.

*Alkali.* A substance which is capable of uniting with acids and destroying their acidity. Potash, soda, etc., are alkalies.

*Alimentary.* Having nourishing qualities, as food.

*Alimentary Canal.* The intestine, by which aliments are conveyed through the body, and the useless parts evacuated.

*Alterative.* A medicine which gradually changes the condition of the system, restoring healthy functions without producing sensible increase of the evacuations.

*Alternate.* When branches and leaves issue singly from opposite sides of the stem, in regular order, first on one side of the stem and then on the other, they are said to be alternate.

*Alveola.* The socket in a jaw in which a tooth is fixed.

*Alvine.* Pertaining to the intestines.

*Amaurosis.* A loss or decay of sight, without any visible defect in the eye, except an immovable pupil.

*Amenorrhea.* An obstruction of the menstrual discharges.

*Ament.* Flowers on chaffy scales and arranged on slender stalks.

*Amplexicaulis.* The basis, clasping the stem.

*Amputation.* The act or operation of cutting off a limb, or other part of the body.

*Anasarca.* Dropsy of the skin and flesh.

*Anastomose.* To communicate with each other.

*Anchylosis.* Stiffness of a joint.

*Aneurism.* A soft, pulsating tumor, arising from the rupture of the coats of an artery.

*Angina Pectoris.* A peculiar, nervous affection of the chest.

*Angina Tonsillaris.* Inflammation of the tonsils.

*Angina Trachealis.* Inflammation of the wind-pipe, or croup.

*Annual.* Yearly. An annual plant grows from the seed to perfection and dies in one season.

*Annulated.* Surrounded by rings.

*Anodyne.* Any medicine which allays pain and procures sleep.

*Antacid.* A substance to counteract acids, as an alkali.

*Anthelmintic.* A worm-destroyer; a worm medicine.

*Antibilious.* Counteraction of bilious complaints.
*Antidote.* A protective against or remedy for poison, or any thing noxious taken into the stomach, or any disease.
*Antidysenteric.* A remedy for dysentery.
*Antiemetic.* A remedy to check or allay vomiting.
*Antilithics.* A medicine to prevent or remove urinary calculi or gravel.
*Antimorbific.* Any thing to prevent or remove disease.
*Antiscorbutic.* A remedy for the scurvy.
*Antiseptic.* That which resists or removes putrefaction or mortification.
*Antispasmodic.* That which relieves spasms, cramps, and convulsions.
*Antisyphilitic.* Remedy against syphilis, or the venereal disease.
*Aperient.* A gentle purgative or laxative.
*Apex.* The top or summit; the termination of any part of a plant.
*Aroma.* The fragrance of plants and other substances, experienced by an agreeable smell.
*Aromatic.* A fragrant, spicy plant, drug, or medicine.
*Arthroida.* A joint movable in every direction.
*Articulated.* Having joints.
*Ascarides.* Pin worms, or thread worms, always found in the lower portion of the bowels, or anus.
*Ascites.* Dropsy of the belly.
*Assimilation.* The conversion of food into the fluid or solid substances of the body.
*Asthmatic.* A person troubled with asthma, or a difficulty of breathing.
*Astringent.* Binding; contracting; strengthening; opposed to laxative.
*Atony.* Debility; want of tone; defect of muscular power.
*Atrophy.* A wasting of flesh and loss of strength, without any sensible cause.
*Axillary.* Pertaining to the arm-pit.
*Axillary Glands*, situated in the arm-pit, secrete a fluid of peculiar odor, which stains linen, and destroys the color of clothing.

*Balsamic.* Medicines employed for healing purposes.
*Belching.* Ejecting wind from the stomach.
*Biennial.* In botany, continuing for two years, and then perishing, as plants whose roots and leaves are formed the first year, and which produce fruit the second.
*Bifurcation.* Division into two branches.

*Biternate.* Doubly ternate, or having six leaves on the leaf stalk.
*Bract.* A small leaf growing near the flower, and differing in form and color from the other leaves.
*Bronchial.* Belonging to the ramifications of the wind-pipe in the lungs.
*Bulbous.* Round, or roundish.

*Cachexia.* A bad condition of the body; where the fluids and solids are vitiated, without fever or nervous disease.
*Cadaverous.* Having the appearance or color of a dead human body; wan; ghastly; pale; like unto death.
*Calculi.* The gravel and stone formed in any part of the body, as the bladder and kidneys.
*Callous.* Hard, or hardened; as an ulcer.
*Callus.* Bony matter, which forms about fractures.
*Caloric.* The element of heat.
*Calyx.* The outer covering of a flower.
*Campanulate.* Bell-shaped.
*Capillary.* Resembling a hair. A fine vessel.
*Capsule.* The seed-vessel of a plant.
*Carbon.* Charcoal.
*Carbonic Acid Gas.* A combination of two parts of oxygen with one part of carbon.
*Carminative.* A medicine which allays pain, and expels wind from the stomach and bowels.
*Cartilage.* Gristle; a substance similar to, but softer than bone.
*Catamenia.* The monthly evacuations of females; menses.
*Cataplasm.* A poultice.
*Cathartic.* A purgative; a medicine that cleanses the bowels.
*Catheter.* A tubular instrument for drawing off the urine.
*Caudex.* The stock which proceeds from a seed, one part forming the body above ground, and the other the main root below.
*Caustic.* Any substance which burns or corrodes the part of living animals to which it is applied.
*Cautery.* A burning, searing, or corroding, any part of an animal body.
*Cellular.* Consisting of, or containing cells.
*Cerebellum.* The hinder and lower part of the brain; the lesser brain.
*Cerebrum.* The front and larger part of the brain.
*Cespitose.* Growing in tufts.
*Cespitous.* Pertaining to turf; turfy.
*Chancre.* A venereal ulcer or sore.
*Choleric.* Easily irritated.

*Cuticle.* The scarf-skin, or outer skin.
*Chronic.* Continuing a long time; inveterate; the opposite of acute.
*Cicatrix.* A scar remaining after a wound.
*Clyster.* An injection; a liquid substance thrown into the lower intestines.
*Coagulation.* Changing from a fluid to a fixed state.
*Coalesce.* To grow together; to unite.
*Colliquative.* Weakening, as sweat; applied to excessive evacuations, which reduce the strength and substance of the body.
*Coma, or Comatose.* Lethargy; strongly disposed to sleep.
*Combustion.* Burning with a flame.
*Concave.* Hollow. A concave leaf is one whose edge stands above the disk.
*Concrete.* A compound; a united mass.
*Confluent.* Flowing together; meeting in their course.
*Congenital.* Begotten or born together.
*Conglobate.* Formed into a ball.
*Connate.* United in origin; united into one body.
*Constipation.* Obstruction and hardness of the contents of the intestinal canal.
*Constriction.* A contraction, or drawn together.*
*Contagious.* Catching, or that may be communicated.
*Contusion.* A bruise.
*Convalescent.* Recovering health and strength after sickness or debility.
*Convoluted.* Rolled together, or one part on another.
*Cordate.* Having the form or shape of a heart.
*Cordial.* Any medicine which increases the strength and raises the spirits when depressed.
*Coriaceous.* Tough or stiff, like leather.
*Corolla.* The inner covering of a flower.
*Corpse.* The dead body of a human being.
*Corroborant.* A medicine that strengthens the human body when weak.
*Corrosive.* That which has the quality of eating or wearing gradually.
*Corrosive Sublimate.* An acrid poison of great virulence.
*Cortex.* The bark of a tree or plant.
*Corymb.* A cluster of flowers at the top of a plant, forming an even, flat surface.
*Cranium.* The skull.
*Crassamentum.* The thick, red part of the blood.
*Crepitas.* A sharp, abrupt sound.
*Cuneiform.* Having the shape or form of a wedge.
*Cutaneous.* Belonging to the skin.

*Decarbonize.* To deprive of carbon, or coal.
*Decoction.* Any medicine made by boiling a substance in water to extract its virtue.
*Delirium.* Disorder of the intellect; wildness or wandering of the mind.
*Demulcent.* A mucilaginous medicine, which sheathes the tender and raw surfaces of diseased parts.
*Deobstruent.* Any medicine which removes obstructions, and opens the natural passages of the fluids of the body.
*Depletion.* Blood-letting.
*Depuration.* The cleansing from impure matter.
*Derm.* The natural covering of an animal, or skin.
*Detergent.* A medicine that cleanses the vessels or skin from offending matter.
*Diagnosis.* The distinction of one disease from another by its symptoms.
*Diagnostics.* The symptoms by which a disease is distinguished.
*Diaphoresis.* Increased perspiration, or sweat.
*Diaphoretic.* Sweating; any medicine which produces sweating.
*Diaphragm.* The midriff, or muscular division between the chest and belly.
*Diarrhea.* A morbidly-frequent evacuation of the intestines.
*Diathesis.* The disposition of the body, good or bad.
*Dichotomous.* Regularly divided by pairs from top to bottom.
*Digest.* To dissolve in the stomach; or, in medicine, to make a tincture.
*Digitate.* Divided, like fingers.
*Diluent.* That which thins, weakens, or reduces the strength of liquids.
*Diluting.* Weakening.
*Discuss.* To disperse, or scatter.
*Discutient.* A medicine which scatters a swelling or tumor, or any coagulated fluid or body.
*Diuretic.* A medicine which increases the flow of the urine.
*Dolor.* Pain.
*Drastic.* Powerful, efficacious.
*Duodenum.* The first of the small intestines.

*Efflorescence.* Eruptions, or a redness of the skin, as in measles, small-pox, etc.
*Effluvia.* Exhalations from substances, as from flowers, or from putrid matter.
*Electuary.* Medicine composed of sugar or honey, and some powder, or other ingredient.
*Eliminating.* Discharging, or throwing off.

*Emetic.* Any medicine which produces vomiting.
*Emaciation.* Gradual wasting of the flesh, leanness.
*Emesis.* A vomiting.
*Emmenagogue.* A medicine which promotes the menstrual discharges.
*Emollient.* A softening application which allays irritation.
*Emulsion.* A soft, milk-like remedy, as oil and water mixed with mucilage or sugar.
*Enema.* An injection.
*Enteritis.* An inflammation of the intestines.
*Entozoa.* Intestinal worms; living in some part of an animal body.
*Epidemic.* A prevalent disease.
*Epidermis.* The outer skin.
*Epigastric.* Pertaining to the upper and anterior portion of the abdomen.
*Epileptic.* Affected with epilepsy, or the falling sickness.
*Epispastic.* An application for blistering.
*Erosion.* The act or operation of eating away.
*Errhine.* A medicine for snuffing up the nose to promote the discharge of mucous.
*Eructation.* The act of belching forth wind from the stomach through the mouth.
*Eruption.* A breaking out of humors on the skin.
*Escharotic.* Caustic; an application which sears or destroys the flesh.
*Evacuant.* A medicine which promotes the secretions and excretions.
*Evacuate.* To empty, to discharge from the bowels.
*Exacerbation.* An increase of violence in a disease.
*Exanthema.* Such eruptive diseases as are accompanied by fevers.
*Excitant.* A stimulant.
*Excoriate.* To gall, to wear off or remove the skin in any way.
*Excrescence.* A preternatural protuberance; as a wart.
*Excretion.* Useless matter thrown off from the system.
*Exotic.* Introduced from a foreign country.
*Expectorant.* Any medicine which promotes the discharge of phlegm, or matter, from the lungs.
*Expectoration.* The act of discharging phlegm by coughing or spitting.
*Expiration.* The act of throwing out the air from the lungs, as in breathing.
*Extravasation.* Effusion; the act of forcing or letting out of its containing vessels.
*Exudation.* A sweating.

*Fæces.* Excrement; the discharge from the bowels at stool.
*Fauces.* The back part of the mouth.
*Febrifuge.* Medicines that drive away fever, producing sweat.
*Febrile.* Indicating fever, or pertaining to fever.
*Fetid.* Having a strong or offensive smell.
*Fetus.* The child while in the womb.
*Fiber.* A fine, slender substance, which constitutes a part of the frame of animals; a thread.
*Fibril.* The branch of a fiber; a very slender thread.
*Filament.* A thread; a fiber.
*Filter.* A strainer.
*Filtration.* Straining; the separation of a liquid from the undissolved particles floating in it.
*Fistula.* A deep, narrow, crooked ulcer.
*Flaccid.* Soft and weak; lax, limber.
*Flatulency.* Wind in the stomach and intestines, causing uneasiness, and often belchings.
*Flexible.* Not stiff; yielding to pressure.
*Flush.* A sudden flow of blood to the cheeks or face.
*Flux.* An unusual discharge from the bowels.
*Fomentation.* Bathing by means of flannels, dipped in hot water or medicated liquid.
*Formula.* A prescription.
*Fundament.* The seat; the terminating part of the large intestines.
*Fungus.* A spongy excrescence, as proud flesh.

*Gangrene.* Mortification of living flesh.
*Gargle.* A wash for the mouth and throat.
*Gastric.* Belonging to the stomach.
*Gland.* A soft, fleshy organ, for the secretion of fluids, or to modify fluids which pass through them.
*Gluteus.* The large, thick muscle on which we sit.

*Hectic.* Habitual; an exasperating and remitting fever, with chills, heat and sweat.
*Hematosis.* A morbid quantity of blood.
*Hemoptysis.* A spitting of blood.
*Hemorrhage.* A flux, or discharge of blood, as from the nose, lungs, etc.
*Hemorrhoids.* The piles.
*Hepatic.* Pertaining to the liver.
*Herbaceous.* Pertaining to herbs.
*Hereditary.* That has descended from a parent.
*Herpes.* An eruption of the skin; tetters, erysipelas, ringworm, etc.

*Hernia.* A rupture and protrusion of some part of the abdomen.

*Hydragogue.* A purgative that causes a watery discharge from the bowels.

*Hydrogen.* A constituent of water, being one-ninth-part.

*Hydrogen Gas.* An aeriform fluid, the lightest body known. It is fatal to animal life.

*Hydrophobia.* A dread of water; the rabid qualities of a mad dog.

*Hygiene.* The art of restoring or preserving the health without recourse to medicine.

*Hypochondriac.* A person afflicted with debility, lowness of spirits, or melancholy—or, in other words, with the blues.

*Hysterical.* Troubled with fits, or nervous affections.

*Idiopathy.* A morbid condition not preceded by any other disease.

*Idiosyncrasy.* Peculiarity of constitution or temperament; peculiarly susceptible of certain extraneous influences—and, hence, liable to certain diseases which others would escape from.

*Ileum.* The lower part of the small intestines.

*Incrassation.* Thickening.

*Incubus.* The nightmare.

*Indigenous.* Native.

*Indurated.* Hardened.

*Infection.* Communication of disease from one to another; contagion.

*Inflammation.* Redness and swelling of any part of the body, with heat, pain, and symptoms of fever.

*Inflated.* Filled or swelled with air.

*Infusion.* Medicine prepared by steeping, either in cold or hot water.

*Ingestion.* Throwing into the stomach.

*Injection.* A liquid medicine thrown into the body by a syringe, or pipe; a clyster.

*Inoculation.* Communicating a disease to a person in health by inserting contagious matter in his skin, or flesh.

*Inspiration.* Drawing or inhaling air into the lungs.

*Inspissation.* Rendering a fluid substance thicker by evaporation.

*Integument.* The skin, or a membrane that invests a particular part.

*Intermittent.* Ceasing at intervals.

*Lanceolate.* Oblong, and gradually tapering toward the outer extremity.

## GLOSSARY.

*Larynx.* The upper part of the wind-pipe.
*Laxative.* A gentle purge; a medicine that loosens the bowels.
*Lethargy.* Unusual or excessive sleepiness.
*Leuchorrhea.* The whites.
*Lesion.* A hurt, or wound.
*Liniment.* A species of soft ointment.
*Lithontriptics.* Solvents of stone in the bladder.
*Lithotomy.* The cutting for stone in the bladder.
*Lochial.* Pertaining to discharges from the womb after childbirth.
*Lumbago.* A pain in the loins, or small of the back.
*Lumbar.* Pertaining to the loins.

*Maceration.* To dissolve, or soften with water.
*Malaria.* Bad air; air which tends to produce disease.
*Manna.* A laxative medicine, obtained from the flowering ash.
*Membrane.* A thin, white, flexible skin, formed of fibers, and covering some part of the body.
*Menses.* The monthly discharges of females.
*Menstrual.* Monthly; occurring once a month.
*Menstruum.* A dissolvent; any liquid used to extract the medical virtue from solid substances.
*Metastasis.* A removal of a disease from one part to another.
*Miasma.* Malaria; infected atmosphere, noxious to health.
*Morbid.* Diseased; not sound or healthful.
*Morbific.* Causing disease.
*Mucilage.* A slimy, ropy, fluid substance.
*Mucous.* A sticky, tenacious fluid, secreted by the mucous membrane.
*Muscles.* The organs of motion. They constitute the flesh.

*Narcotic.* A stupefying, sleep-producing medicine, often administered to relieve pain.
*Nausea.* Any sickness accompanied with an inclination to vomit.
*Nephritic.* A medicine for curing diseases of the kidneys.
*Nervine.* A medicine that operates on the nerves.
*Normal.* Regular, natural.
*Nutritious.* Nourishing.

*Oblong.* Longer than broad.
*Obtuse.* Dull; not acute.
*Omentum.* The caul, or covering of the bowels.
*Opthalmia.* Inflammation of the eyes.
*Ossify.* To change flesh, or other soft matter, into a hard, bony substance.
*Oval.* Egg-shaped.

*Oxygen.* A constituent part (being about one-fifth) of atmospheric air.

*Palpitation.* A beating of the heart; sometimes, a violent beating of the same, caused by fear, etc.
*Panacea.* A universal medicine.
*Paralysis.* A loss of the power of motion in a part of the system.
*Paralytic.* Affected with, or inclined to, palsy.
*Paroxysm.* A fit of any disease.
*Pathology.* The doctrine of the causes, symptoms, and nature of disease.
*Pectoral.* Pertaining to the breast. Medicine for the cure of breast and lung complaints.
*Peduncle.* The stem that supports the flower and fruit of a plant.
*Perennial.* Continuing more than two years; perpetual.
*Pericardium.* A membrane inclosing the heart.
*Permeate.* To pass through the pores.
*Perspiration.* Insensible evacuation of the fluids of the body through the pores of the skin; also, the matter thus discharged.
*Petiole.* A leaf-stalk.
*Petechiæ.* Purple spots on the skin in malignant fevers.
*Pinnate.* A pinnate leaf is a species of a compound leaf.
*Plethoric.* Fullness, or excess of blood.
*Pleura.* A thin membrane, which lines the inside of the chest and invests the lungs.
*Pneumonia.* An inflammation of the lungs.
*Polypus.* A pear-shaped tumor.
*Prolapsus.* A falling down, or falling out, of some part of the body.
*Prophylactic.* A medicine to prevent disease.
*Pubescent.* Covered with down, or with very fine, short hairs.
*Pulmonary.* Pertaining to, or affecting the lungs.
*Pulp.* A soft mass.
*Pungent.* Sharp, piercing, biting, stimulating.
*Purgative.* A medicine that evacuates the bowels.
*Purulent.* Consisting of pus, or matter.
*Pus.* The yellowish, white matter in ulcers, wounds, and sores.
*Pustules.* Pimples.
*Putrescent.* Becoming putrid, or rotten.
*Pyrosis.* A peculiar disease of the stomach, commonly called water-brash.

*Rectum.* The last part of the large intestines.

*Refrigerant.* A cooling medicine.
*Regimen.* The regulation of diet, in order to preserve or restore health.
*Resolvent.* A medicine for driving away inflammation, and to prevent tumors from coming to a head.
*Restorative.* A medicine for restoring vigor and strength.
*Resuscitate.* To recover from apparent death.
*Reticulated.* Like net-work.
*Rigid.* Stiff; not easily bent.
*Rubefacient.* An application which produces redness of the skin.
*Rubific.* Making red.

*Saccharine.* Having the qualities of sugar.
*Saliva.* Spit, or spittle. It serves to moisten the mouth and tongue, and also the food.
*Salivation.* The act of increasing the secretion of saliva.
*Sanative.* Healing, or tending to heal.
*Sanguine.* Having the color of, or abounding with, blood.
*Scirrhous.* Hard: knotty.
*Scorbutic.* Pertaining to, or partaking of, the nature of scurvy.
*Scrotum.* The pouch, or bag, which contains the testicles.
*Secretion.* The act of producing from the blood substances different from the blood itself, or from any of its constituents, as bile, saliva, mucous, &c.; also, the matter secreted.
*Sedative.* A quieting, soothing medicine, which allays irritation, and assuages pain.
*Sedentary.* Accustomed to, or requiring much, sitting; inactive.
*Seminal.* Pertaining to, or contained in, seed.
*Septic.* A promotive of putrefaction.
*Serous.* Thin, watery; like whey.
*Serum.* The watery parts of blood, or of milk.
*Sinaplasm.* A mustard plaster.
*Sinew.* That which unites a muscle to a bone.
*Sialagogue.* Medicines which excite an increased flow of saliva.
*Slough.* To separate from the sound flesh; as the matter formed on a sore.
*Solution.* A liquid in which a solid substance has been dissolved.
*Solvent.* Having the power of dissolving solid substances.
*Spasm.* A violent but brief contraction of the muscles, or fibers.
*Spasmodic.* Consisting in, or relating to, spasms.

*Spleen.* The milt.
*Stimulant.* An exciting agent.
*Stomachic.* A strengthening medicine for the stomach, exciting its action.
*Stool.* A discharge from the bowels.
*Strangury.* A painful and difficult discharge of the urine.
*Stricture.* A morbid contraction of any passage of the body.
*Styptic.* A medicine which coagulates the blood and stops bleeding.
*Sudorific.* A medicine that produces sweat.
*Suppurate.* To form purulent matter, or pus; as a boil.
*Suture.* The peculiar joint uniting the bones of the skull.
*Syncope.* A fainting, or swooning.
*Syphilitic.* Pertaining to the venereal disease, or pox.

*Tendon.* A bunch of fibers attaching a muscle to a bone.
*Tenesmus.* A distressing pressure, as if the bowels must be discharged immediately.
*Tense, or Tension.* Stretched, or strained; rigid.
*Tepid.* Moderately warm.
*Terminal.* Forming the end; growing at the end of a branch or stem.
*Ternate.* Three leaves together on a leaf-stalk.
*Tertian.* An intermittent fever or disease, in which the fits or paroxysms return every other day.
*Tincture.* Medicine dissolved in alcohol, or proof spirits.
*Thorax.* The cavity of the chest.
*Tomentose.* Downy, nappy; covered with the finest hairs, or down.
*Trachea.* The wind-pipe, or breathing passage.
*Translated.* Removed from one place to another.
*Transude.* To pass through pores or interstices.
*Triennial.* Lasting three years.
*Tubercle.* A pimple; a swelling, or tumor.
*Tuberous.* Consisting of roundish, fleshy bodies, as potatoes.
*Tumefaction.* The act of swelling, or forming a tumor.
*Tumor.* A distension or enlargement of any part of the body; a swelling.
*Tunic.* A membrane that covers or composes some part or organ.
*Typhoid.* Resembling typhus; weak, low.
*Typhus.* A simple, continuous fever, attended with exhaustion, weakness of pulse, and frequently strong propensities to sleep.

*Ulcer.* A sore, discharging pus.
*Umbilic.* The navel; or pertaining to the navel.

*Ureter.* A duct or tube, through which the urine passes from the kidneys to the bladder.
*Urethra.* The canal that receives the urine from the bladder and discharges it.
*Urinary.* Pertaining to urine.
*Urine.* A fluid secreted by the kidneys, and conveyed from the bladder through the urethra and discharged.
*Uterus.* The womb; that part of a female where the child is produced.

*Vaccinate.* To communicate the cow-pox to a person by inserting the vaccine matter in the skin.
*Vaccine.* Derived from cows.
*Vagina.* The canal leading from the external orifice to the womb.
*Varioloid.* A modified variety of small-pox.
*Variolous.* Pertaining to, or designating the, small-pox.
*Venery.* Intercourse of the sexes.
*Vermifuge.* A worm-destroyer; or a medicine to expel worms.
*Vertigo.* Dizziness, or swimming of the head.
*Vesication.* Raising blisters on the skin.
*Vesicle.* A small cavity; a little bladder filled with some humor.
*Virus.* Contagious matter; poison.
*Viscera.* The bowels and internal organs of the body.
*Viscid.* Sticky, tenacious, like glue.
*Vitiate.* To injure; to impair; to spoil.
*Volatile.* Substances which waste away on exposure to the atmosphere.
*Vulnerary.* Medicines used for the cure of wounds.

## Let the Afflicted Read Carefully this Page!

### THE
# Health Reform Institute,
### BATTLE CREEK, MICH.

This Institution was opened for the reception of Patients and Hygienic Boarders, on the 5th of September, 1866. Diseases are here treated on Hygienic Principles, and instruction is imparted both thoretically and practically, to patients and boarders, on the important subject of Preserving Health as well as Recovering from Disease. No Drugs are ever administered. Water, Air, Light, Heat, Food, Sleep, Rest, Recreation, &c., are the only agents resorted to in treating the sick. Vegetables, Grains and Fruits, are the staples in the dietary. The water is pure and soft,—of the best quality. Grounds ample and pleasant. The charges moderate. Full particulars given in Circulars, which will be sent free on application.

Address,     HEALTH INSTITUTE, BATTLE CREEK, MICH.

## THE HEALTH REFORMER.

A Monthly Journal, published at The Health Reform Institute, Battle Creek, Mich. The object of this Journal is to aid in the great work of reforming, as far as possible, the false habits of life so prevalent at the present day. It will aim to teach faithfully and energetically those Rules of Health, by obedience to which, people may secure the largest immunity from sickness and premature death. It will advocate the cure of diseases by the use of NATURE'S OWN REMEDIES. such as Water, Air, Light, Heat, Exercise, Food. Sleep, Recreation, &c. It will conscientiously hold up the light on the best methods, so far as ascertained, of managing healthfully our physical frames. It will be adapted to the wants of all classes of people everywhere, who are interested in the great question of maintaining health by obedience to Nature's Laws; and where such interest does not exist, it will endeavor to create it; for which purpose we wish to give it a wide and indiscriminate circulation. And to make it more specially practical, and adapt it to the more immediate wants of the people, a certain amount of space will be devoted in each number to the answering of questions from correspondents, and giving directions for the use of water, and the home treatment of disease.

In short, we aim to publish a First Class Health Journal, interesting in its variety, valuable in its instructions, and second to none in either literary or mechanical execution. The year's numbers, when bound, will furnish a volume of nearly 200 pages, convenient in size, and filled with the choicest reading matter. We solicit subscriptions from all the friends of the health movement, and ask them to lend their aid in extending the circulation of this Journal. Price $1.00, in advance, per volume of twelve numbers.

☞ Specimen numbers sent free to any address.

Address,     HEALTH REFORMER, BATTLE CREEK, MICH.

# HEALTH IS HAPPINESS!

## ☞ READ AND PRESERVE. ☜

## BOOKS! REFORMER! INSTITUTE!

### The Hygienic Family Physician.

THIS is the title of a work recently published at this Office. As the title suggests, it is a work especially designed for family use. The style in which it is written is such as to render it perfectly intelligible to all classes, as it is quite free from technical terms and phrases which are of such frequent occurrence in nearly all books of this kind which have previously appeared as to render them more or less objectionable. It is, nevertheless, "a complete guide for the preservation of health and the treatment of disease without the use of medicine."

The work is written in four parts. The subjects treated are, in Part I., Health and Hygienic Agents; Part II., Disease and Drugs; Part III., the Bath; Part IV., Diseases and their Treatment. A more minute description of each part is found below. This work is of a thoroughly practical nature, and should be in the hands of every family in the land, as it affords instruction of the most vital importance. Directions for the treatment of disease are so plain and minute that any person of ordinary intelligence with its assistance may successfully treat nine-tenths of all the cases of disease which occur in any neighborhood. The publishers have placed the price so low that the book may be obtained by any one who feels at all in need of such a work.

Published at the *Health Reformer* Office. Cloth, bound, 380 pp. Price, post-paid, $1.00.

---

The following four pamphlets contain the larger portion of the bound work just noticed. They severally correspond with the four parts of the bound volume.

### Good Health, and How to Preserve It.

In this pamphlet is given a brief treatise on the various hygienic agents and conditions which are essential for the preservation of health. Just the thing for a person who wishes to learn how to avoid disease.

Published at the *Health Reformer* Office. Price, post-paid, 10 cents.

### Nature and Cause of Disease, and So-called "Action" of Drugs.

This work is a clear and comprehensive exposition of the nature and true cause of disease, and also exposes the absurdity and falsity of drug medication.

Published at the *Health Reformer* Office. Price, post-paid, 15 cents.

### The Bath: Its Use and Application.

This very valuable work contains a full description of the various baths employed in the hygienic treatment of disease, together with the manner of applying them, and the diseases to which they are severally adapted.

Published at the *Health Reformer* Office. Price, postpaid, 20 cents.

### The Treatment of Disease.

In this most important work may be found an accurate description of the symptoms and proper treatment of more than one hundred diseases, comprising all of those which are susceptible of ordinary home treatment. It is an invaluable work for all who are not professionally educated in the theory and practice of medicine. The only remedies recommended are of course strictly hygienic in their nature, drugs of every description being entirely discarded as curative agents.

Published at the *Health Reformer* Office. Price, postpaid, 35 cents.

---

### The Hygienic System.

#### By R. T. Trall, M. D.

This important work treats upon the Principles of Hygienic Medication—Hygeio-Therapy—The Essential Nature of Disease—The Modus Operandi of Medicine—The Relations of Remedies to Diseases—The Relations of Remedies to the Healthy Organs—The Doctrine of Vitality—The Law of Cure—The Problems of Medical Science. It should be read by the million.

Published at the *Health Reformer* Office. Price, postpaid, 15 cents.

### Health and Diseases of Woman.

#### By R. T. Trall, M. D.

This work treats upon Woman and the Medical Profession—Opium—Alcohol—Tobacco—Drugs—The Race Imperiled—Responsibilities of Parents—American Mothers—Woman's Disadvantages—The Medical Profession *vs.* Woman—Origin of Many Infirmities—Dress and Respiration—Dress and the Sexual Functions—Should Fashionable Women Marry?—Drugging at Puberty—Scientific Druggery—Scanzoni *vs.* Churchill—Dr. Prescott on Druggery—Drugging in Acute Diseases—Prof. Gilman on Puerperal Fever—Drugging During Pregnancy—Drugging During the Lying-in Period—Chronic Drug Disease—the Better Way—Tobacco *vs.* Woman.

It should be in every family, and be read by every woman and every girl in the land.

Published at the *Health Reformer* Office. Price, post-paid, 15 cents.

### Tobacco-Using.

#### By R. T. Trall, M. D.

This is a Philosophical Exposition of the Effects of Tobacco on the Human System. Published at the *Health Reformer* Office. Price, post-paid, 15 cents.

### Science of Human Life.

This is a pamphlet of great value, containing three of the most important of Graham's Lectures on the Science of Human Life. It is published for the benefit of those who may not feel able to purchase the entire work, and contains most of that work which is of practical value to the reading public.

Published at the *Health Reformer* Office. Price, postpaid, 35 cents.

### Hand Book of Health.

This work treats upon Physiology and Hygiene.
Published at the *Health Reformer* Office. Price, postpaid, bound in cloth, 60 cents; in paper cover, 35 cents.

### Cook Book, or Kitchen Guide.

This work comprises recipes for the preparation of hygienic food, directions for canning fruit, &c., together with advice relative to change of diet.

Published at the *Health Reformer* Office. Price, postpaid, 20 cents.

### EXHAUSTED VITALITY;

Or, a Solemn Appeal Relative to Solitary Vice, and the Abuses and Excesses of the Marriage Relation. We do not hesitate to say that this is the best work of the kind now in print in our country. It is gathered chiefly from the writings of the ablest and best writers upon the subject. Of this subject, and this work, the compiler in his preface says :—

"It is disagreeable to call attention to those sins of youth, and the abuses and excesses, even in the married life, which are ruining the souls and bodies of tens of thousands ; especially so, while feelings of great delicacy, relative to the subject, exist in the public mind. But disagreeable though the task may be, facts, terrible facts of every-day observation, fully justify a solemn and faithful warning to all. We would cherish the profoundest respect for the delicate feelings of the truly modest and the really virtuous ; but we confess our want of respect for that false delicacy in many which takes fright at the mention of those vices, in consequence of which, they themselves exhibit evident marks of rapid decay.

"The reader may as well prepare at the first, by laying aside feelings of false delicacy, if he is troubled with them, to be benefited by the painful facts, plainly stated in this work. The real value of the lengthy article on

"CHASTITY"

Cannot be estimated by dollars and cents. Every youth in the land should read it. And not only the youth, but every parent and guardian, should study it well, and be prepared in a proper way to warn those children under their immediate care. And let every mother be stirred by the article under the caption of

"APPEAL TO MOTHERS."

It comes from a mother's heart—from one who has had experience in laboring for the unfortunate victims of secret vice, and is imbued with the importance of the subject. The extracts entitled

"EVILS AND REMEDY,"

Although unvailing many dark pictures, are entitled to consideration as the utterances of one whose extensive study of human nature has qualified him to speak to the point on this important subject."

Published at the *Health Reformer* Office. Price, postpaid, bound in cloth, 60 cents; in paper cover, 30 cents.

## THREE-CENT TRACTS.

The following tracts are offered, post-paid, for three cents each, or two dollars per hundred. This list of tracts will be greatly increased.

**Dyspepsia**: Its Causes, Prevention, and Cure.

**The Dress Reform**: Containing reasons for the most Healthful, most Modest, and most Convenient Style of Woman's Dress.

**The Principles of Health Reform**: Important to those whose minds should be called to first principles.

# THE HEALTH REFORMER.

This is a monthly journal devoted to physical, mental, and moral culture.

## ITS MISSION.

As indicated in the prospectus, its mission is to contribute to the improvement of mankind physically, mentally, and morally. Of the necessity for reform in these particulars, we need not speak; for the alarming evidences of physical degeneracy and disease, mental inefficiency, and moral turpitude, which we see about us on every hand, speak more loudly than can words of the crying need of immediate and thorough reformation.

Progression is the spirit of the times. Social reform, prison reform, civil service reform, and various other reforms, each in its turn, calls for the careful and candid consideration and hearty co-operation of every intelligent man and woman. And very just and appropriate is this demand; for nothing can be more promotive of the interests of society than improvement—progression—*reform*. Without this, stagnation would result, and civilization would soon degenerate into the veriest barbarism. Its value, then, cannot be overestimated; and every truly reformatory movement should receive our most serious and attentive consideration.

As its name would suggest, the *Health Reformer*

is published in the interest of a reformation which has a special bearing upon health; health—physical, mental, and moral. Perfect physical development, clear mental faculties, and acute moral sensibilities, constitute the perfection of manhood or womanhood. Can there be anything more important, then, than a reform which aims to secure these three conditions, which, when attained, will place a person in that state of perfection which will enable him to realize the highest degree of enjoyment possible for man to experience? May we not justly claim that, while the reforms which have been mentioned are of great moment and absorbing interest, they are all eclipsed by the far greater importance of this reform which deals with those principles which underlie the whole superstructure of moral and social life, and which strike at the very root of all the evils which curse society, and rest like a mighty incubus upon the world?

## PLAN OF ACTION.

In order to accomplish the desired object, which has already been set forth, the conductors of the *Reformer* have adopted this as a fundamental principle of action: Physical reform is the basis of all reform. The truth of this principle is evident when we consider,

1. The intimate relation of mind and matter, and the wonderful manner in which the mind is affected by the varying conditions of the body; so that whenever the body suffers from serious injury of any kind, the mind is correspondingly impaired, as is seen in the fever patient raving in the wildness of delirium.

2. The fact that the condition of a person's moral organs depends so largely upon that of the body and mind; as is illustrated by the victim of despair who labors under the impression that his doom is sealed, when his only difficulty is a torpid liver; or the irritable, misanthropic dyspeptic, whose unhappy mental condition is wholly due to a disordered stomach.

In view of these facts, it appears that the most important branch of the work of the *Reformer* is in the direction of physical improvement and reform, since the success of each of the other branches is contingent upon the success of this.

But while constantly aiming at reform, and so contending against adverse and opposing influences, the conductors of the *Reformer* are careful to avoid those extremes into which so many reformers allow themselves, unwittingly, perhaps, to be led. They also ever seek to manifest that liberality of sentiment which is in harmony with the spirit of the present time, when every man is expected and urged to think and form opinions for himself. By so doing, they hope to incite a spirit of investigation, which, when pursued with candor and an unbiased judgment, can hardly fail to convince the reader of the truth of the positions taken.

Those who conduct the *Reformer* endeavor to fill its columns with matter of practical importance and interest to every subscriber. Thorough instruction is given in regard to these two most important subjects,

## HOW TO RECOVER HEALTH, AND HOW TO RETAIN IT,

These subjects being treated by those whose personal experience enables them to speak understandingly. In fact, we put forth every effort to make the *Reformer indispensable to every household*, and of especial interest to that exceedingly large and unfortunate class of individuals who have been brought into the condition of invalids by disease. But the subject of health, proper, by no means receives exclusive attention. Considerable space is each month devoted to general literature, important and interesting discoveries in the arts and sciences, and such other subjects as are of general interest.

## PRESENT PROSPECTS.

Notwithstanding the numerous and almost insurmountable obstacles with which it has been obliged to contend, the *Reformer* has made constant and rapid progress in extending its sphere of usefulness, until it is now established upon a firm and satisfactory basis, being furnished with an able corps of contributors, numbering its patrons by thousands throughout the United States and Territories.

The publishers of this journal are actuated by purely philanthropic motives, and hence offer it at such terms as will enable every person to obtain it who has any degree of interest in the important subjects, How to GET WELL and how to KEEP WELL. Terms, $1.00 a year, in advance. Specimen copies sent free on application. Address, HEALTH REFORMER, *Battle Creek, Mich.*

# THE HEALTH INSTITUTE.

### LOCATION.

THIS model health institution is situated in the most healthful and delightful part of the proverbially neat and enterprising city of Battle Creek, Michigan, an important station on the Michigan Central R. R., about half way between Chicago and Detroit. Several railroads intersect at this point, making it easy of access from all directions.

### GROUNDS.

The grounds are ample, consisting of a site of about twenty acres, a large portion of which is covered with shade, ornamental, and fruit trees. They are also high, overlooking the entire city, and affording a fine view of the landscape for miles around.

The soil is of such a nature that mud is almost entirely unknown, a few hours of sunshine after a rain rendering the walks and roads in and about the grounds so free from dampness that the most delicate invalid may indulge freely in the benefits of out-of-door life and exercise.

In front of the main building, and between it and the road, is a beautiful grove, which extends along the street in each direction from it, some thirty rods, affording a delightful place of resort during the summer months. The grove is also provided with such means of exercise and recreation as are both healthful and entertaining; as croquet grounds, conveniences for gymnastic exercises, etc.

## BUILDINGS.

These comprise a large main building, and seven fine cottages, all situated upon the same site. The main building contains commodious parlors, dining halls, bath and movement rooms, etc., etc., while the other buildings are fitted up as private apartments for patients. By this means are secured that quiet and retirement which are of such paramount importance to the invalid, and which cannot be obtained in an institution where scores of suffering individuals are crowded together under one roof.

## ROOMS

Are large and well ventilated, and are furnished much better than in any other institution of the kind, thus affording the patient all the luxuries and comforts which he enjoys at home, and many more.

## PLAN OF TREATMENT.

At this institution diseases are treated on strictly hygienic principles; that is, only those remedies are employed which will assist nature in her healing work, and will in no way endanger the life or constitution of the patient. Drugs and poisons of every description are entirely discarded as curative agents; but all known means of restoring health are constantly employed, poisons alone being excluded from our materia medica.

## OUR REMEDIES

Then are Light, Water, Air, Electricity, Exercise, Cheerfulness, Rest, Sleep, Proper Clothing, Proper Food, and, in fact, all Hygienic and Sanitary Agents.

## OUR PHYSICIANS.

The medical faculty of the institution is composed of an adequate number of conscientious, watchful and efficient physicians, who give personal and unremitting care and attention to their patients, anticipating, as far as possible, their wants, carefully studying their cases, and applying every means in their power to restore them to health.

## OUR FACILITIES.

Very few institutions are provided with conveniences and advantages equal to ours. Our bath rooms are both capacious and convenient, and are furnished with an inexhaustible supply of pure, soft water. Several rooms are also prepared especially for the administration of the Sun-Bath.

## SPECIAL ADVANTAGES.

In addition to the appliances usually employed in such institutions, we make use of the Hot-Air Bath (which possesses all the virtues of the Turkish-Bath, while avoiding its evils), the much-renowned Electric or Electro-Thermal-Bath, the Lift Cure, and the celebrated Swedish Movement Cure, which are so successful in many cases which cannot be reached by other means.

## DIET.

While we reject from our dietary those pernicious drinks and condiments which are the potent agents in bringing thousands to untimely graves, we take care to supply our table with an abundance of nutritious and palatable food, consisting of fruits, grains, and vegetables. We do not enforce, however, a radical and immediate change from old habits, but give the patient time to accommodate himself to the new diet.

## OUR SUCCESS.

The class of individuals who seek aid at our institution is very largely composed of those who are afflicted with chronic diseases, and who have been drugged and poisoned until their vitality has become well-nigh exhausted, and they are given up by their friends and medical advisers to die. Under these circumstances, they come to us as a final resort, and, thanks to a true and potent system of treatment, this last hope is seldom disappointed. Among the hundreds who have thus come to us and found relief from their ills and pains, during the eight years since the establishment of this institution, the following cases, here briefly reported, have been treated within the last few months:—

## CONSUMPTION.

Many cases might be cited, and references given, in which this most insidious and hopeless of all diseases has been robbed of its victims and a new lease of life given them by a few months' stay with us.

## DYSPEPSIA.

Hundreds have come to us afflicted with this most deplorable disease in its most aggravated forms, and, after staying a proper time, have returned to their friends relieved of their sufferings.

## PARALYSIS.

Even this formidable disease is, in many cases, treated with entire success, the use of paralyzed organs being wholly restored.

## DROPSY.

In one case, the patient came to the Institute after having been given up to die by friends and physicians. He had been tapped many times, as the accumulation of fluid was so rapid that respiration was rendered extremely difficult in a few days. Cured in a few months, and reports himself still in good health.

## SCROFULA.

Many cases of scrofula, often complicated with dyspepsia, affections of the lungs, etc., have been treated with marked success. In one case, the patient had several large tumors, one nearly as large as an ordinary bowl. After a few weeks' treatment, nature began the curative work of absorption, thus effecting a cure. This case had been considered entirely hopeless.

But space will not allow further description of the desperate cases which have received treatment and restoration at this institution; but we may add that equally good success has attended the treatment of Asthma, Kidney Difficulties (of the worst forms), Chronic Diarrhea, Chronic Congestion of the Brain, Cancer, Palpitation of the Heart, Rheumatism, Neuralgia, Epilepsy, Bronchitis, Piles, Ulceration of Bowels, Catarrh of Bladder and Bowels, Constipation (in some cases without a natural passage for many years), Spermatorrhea, and, in fact, Chronic Diseases of all kinds.

The most flattering success has attended the treatment of Uterine Difficulties, and all other Diseases of Women, which receive special attention.

## ACUTE DISEASES.

Our mode of treatment is specially adapted to this class of diseases, meeting with the most uniform success with Fevers and Inflammations of every type and form, all Eruptive Diseases, etc., etc.

To the sick, we say, Do not delay seeking our assistance until your case is hopeless. Write at once for our Circular, which will be sent free on application.

Address,         HEALTH INSTITUTE,
*Battle Creek, Mich.*

www.ingramcontent.com/pod-product-compliance
Lightning Source LLC
Chambersburg PA
CBHW031737230426
43669CB00007B/382